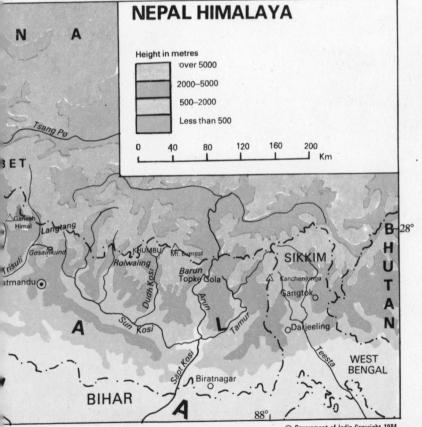

Areas used in recording Geographical distribution.

Extent of this boo[k]

Km 0

D1336767

TIBET

SOUTH EAST TIBET

SIKKIM

EAST BHUTAN ARUNACHAL PRADESH

N A

L

SOUTH

WEST

CHINA

BURMA

BANGLADESH

NEPAL HIMALAYA

Height in metres

over 5000

2000–5000

500–2000

Less than 500

Km 0   40   80   120   160   200

N A

Tsang Po

ET

△Ganesh Himal

Langtang

Gosainkund

Trisuli

atmandu ◉

KHUMBU

Rolwaling

Dudh Kosi

Sun Kosi

Mt. Everest

Barun

Topke Gola

Arun

L Tamur

Sapt Kosi

Biratnagar

BIHAR

A

SIKKIM

△Kanchenjunga

Gangtok ◉

Darjeeling ○

Teesta

B 28°

H

U

T

A

N

WEST BENGAL

88°

FLOWERS OF THE HIMALAYA

A Supplement

ADAM STAINTON

DELHI
OXFORD UNIVERSITY PRESS
BOMBAY CALCUTTA MADRAS
1988

Oxford University Press, Walton Street, Oxford OX2 6DP

New York   Toronto
Delhi   Bombay   Calcutta   Madras   Karachi
Petaling Jaya   Singapore   Hong Kong   Tokyo
Nairobi   Dar es Salaam
Melbourne   Auckland

and associates in
Berlin   Ibadan

ISBN 0-19-217756-7

Text typeset at Sri Aurobindo Ashram Press, Pondicherry
and printed at Rekha Printers (Pvt.) Ltd., New Delhi 110020
Plates printed at Indraprastha Press (CBT), New Delhi 110001
Published by S. K. Mookerjee, Oxford University Press
YMCA Library Building, Jai Singh Road, New Delhi 110001

Contents

Maps

Acknowledgements

Several kind friends have lent photographs for inclusion in this book. I thank them all for their generosity.

Dr C. Grey Wilson 1, 33, 60, 148, 150, 153, 154, 235, 246, 497, 498, 517, 519.

Mr R. McBeath 174.

Mrs L. Polunin (widow of the late Oleg Polunin) 120, 210, 218, 234, 251, 278, 279, 296, 341, 356, 358, 424, 431, 447, 474, 554, 557, 562, 575, 581.

Royal Botanic Garden Edinburgh 290.

Mr A. D. Schilling 20, 24, 28, 64, 65, 84, 86, 98, 131, 149, 180, 227, 247, 255, 291, 294, 327, 366, 395, 435, 436, 440, 444, 451, 455, 467, 500, 522, 525, 530, 532, 543, 544, 577.

Mr G. F. Smith 3, 14, 30, 32, 44, 48, 159, 161, 162, 163, 164, 166, 167, 168, 263, 266, 287, 295, 297, 300, 301, 305, 311, 312, 313, 315, 348, 349, 350, 351, 352, 354, 381, 445.

Mr J. C. H. Templer 298.

Additional Bibliography

A. J. C. Grierson & D. G. Long (1983-), *Flora of Bhutan*, Roy. Bot. Gard. Edin. This flora also includes species recorded from Sikkim. Volume 1, parts 1 & 2, has now been published, and the whole work is planned for completion within the next few years. When finished the work is likely to prove invaluable for all serious students of the flora of the more eastern parts of our area.

Adrian & Jimmie Storrs (1984), *Discovering Trees in Nepal and the Himalayas*, Sayayogi Press, Kathmandu. This useful book describes 170 species of trees and shrubs commonly seen in Nepal. It is illustrated with black and white photographs.

M. L. Banerji & P. Pradhan (1984), *The Orchids of Nepal Himalaya*, J. Cramer, Vaduz. This large folio volume is intended for the study rather than the field. It covers all the orchid species contained in the herbarium of the Botanical Survey of Nepal, Kathmandu. Each species is described and illustrated by drawings.

Introduction

This book is intended as a Supplement to *Flowers of the Himalaya*. Some explanation is due as to why, so comparatively soon after the publication of that book, a Supplement should be thought desirable.

Flowers of the Himalaya was published in 1984, but its inception goes back many years earlier. The late Oleg Polunin and I first discussed in 1956 the lack of any book to help us identify Himalayan plants in the field, and the possibility of one day producing such a book ourselves. The idea then was no more than a pipe-dream, for apart from our own slender knowledge at that time of the Himalayan flora, the botanical exploration of Nepal was not sufficiently advanced to make such a book possible. However, as the years went by further collecting took place in the field, and a massive amount of work was done on these collections both at the British Museum (Natural History) in London, and in Japan. This work ultimately culminated in 1978–82 in the publication of the *Enumeration of the Flowering Plants of Nepal*.

As the preparation of the *Enumeration* advanced, Polunin and I returned to our idea of a Himalayan field-guide. We already possessed a number of plant-photos taken on previous journeys, and during the next few years we added greatly to their number. With so rich a flora it would have been possible to continue almost indefinitely to add to our photo-collection, but practical considerations had to prevail. In 1979 we decided that the coverage was adequate, and in that year the photographic lay-out of *Flowers of the Himalaya* was completed.

While working on this lay-out it became evident that there would not be space to include all the photos which we had already accumulated, and after 1979 I continued to travel and photograph extensively in the Himalaya until 1985. As a result there are now available almost as many new photos as were included in *Flowers of the Himalaya*. These photos are not readily accessible to the public in the form of transparencies, and it seems worthwhile to reproduce them in printed form in order to make as full a record as possible of the Himalayan flora. In some respects it would have been preferable to incorporate both the old and the new material in a new edition of the original book, but *Flowers of the Himalaya* in its present form is already rather large for field use, and the inclusion of new material would make its size excessive. More decisively, considerations of expense preclude any such solution, for to rearrange the photos and remake all the colour separations would be prohibitively costly. It seems best therefore to present the new material in the form of a Supplement.

The new material falls into three main categories: species which were not included in the previous book; species described previously, but not illustrated; species previously described, and illustrated by line-drawings but not by photos. In the latter case it seems worthwhile to provide an additional photographic record, for though botanists may well prefer drawings as a means of identification, the general public probably find it easier to relate to a coloured photo.

I must ask my readers, who are likely to come from more than one part of the world, for a degree of mutual forbearance on the question of which species have been thought worthy of inclusion. Indians may well be surprised to find included in a Himalayan flora species which are common in the Indian plains. Likewise, Europeans may be surprised at the inclusion of some common European species. Plants which are so well-known as to be common-place to the one group may well be unfamiliar to the other.

A considerable number of subtropical species have been included in this Supplement. Nowadays many people visit the Himalaya outside the summer months, and both they and those who live and work in places such as Kathmandu all the year round are in fact much more likely to see plants in flower at subtropical rather than alpine altitudes. A number of exotics have also been included, many of them of American origin. These plants are often showy, and experience has shown that many visitors suppose them to be part of the indigenous flora.

It has been assumed that those who acquire the Supplement will already possess the original work. Plants previously described have therefore not been described again, but reference is made to the appropriate page of *Flowers of the Himalaya* (abbreviated in the text as F.H.). The disadvantage of having to refer from one book to the other seems preferable to burdening a book intended for field use with additional pages which are not strictly necessary. For the same reasons, readers are referred to the glossary in the original book.

Undoubtedly there is considerable resistance amongst the general public to the use of Latin names, despite the indisputable fact that this is the only universally valid way of referring to a plant. When describing the Himalayan flora English names can be applicable only where the species also occurs in Britain, or where it is so well-known as to have acquired an English name even though it does not grow there. After the publication of *Flowers of the Himalaya* some regrets were expressed that no vernacular names had been included. The difficulty here is that more than one language is involved, not only Hindi and Nepali, but also the various Tibetan dialects, the Nepali caste languages, Kashmiri, and doubtless other variants as well. And even amongst speakers of the same language the name used for a plant may well differ from one area to another. To include vernacular names in the text would therefore be a task to tax the skills of a trained philologist.

Very recently it has been learnt that the plant illustrated on Pl. 103 and there captioned as 467 *Euphorbia* aff. *sikkimensis*, will shortly be published as a new species, *Euphorbia schillingii* A. Radcliffe-Smith. The principal distinction is that its fruits are warty, whereas the fruits of *Euphorbia sikkimensis* are smooth.

RANUNCULACEAE   Buttercup Family

1 **Aconitum bhedingense** Lauener
Perennial with erect stem to 40 cm, white-hairy above. Leaves 3–lobed, hairless or downy beneath, sometimes with bulbils in axils; lobes cut into narrow toothed segments. Flowers mauvish-blue or pale yellow, 2–3 cm, in loose erect clusters. Sepals hairy; uppermost sepal helmet-shaped, with acute sometimes reflexed tip; lateral sepals obovate; lower sepals elliptic-oblong, deflexed. Follicles hairy.
    Endemic to E. Nepal. 3000–3900 m. Sep. **Pl. 3.**

2 **Aconitum gammiei**   See F.H. p. 5. **Pl. 1.**

3 **Aconitum naviculare** (Brühl) Stapf
Biennial with tuberous roots and erect or ascending stem 12–30 cm. Leaves mostly basal, rounded-kidney-shaped, palmately divided into 3–5 segments; segments obovate, with 2–3 entire or toothed lobes. Stem-leaves 1–2, more or less stalkless, deeply divided into 3 segments; segments with 3–4 oblong-linear lobes. Flowers 1–2, on stalks with curved or spreading hairs. Sepals whitish or violet, flushed blue-purple, with prominent darker veins; uppermost sepal boat-shaped, obliquely erect, 16–20 mm long; lateral sepals oblique, orbicular-ovate; lower sepals elliptic blunt, 8–12 mm.
    W. Nepal to Bhutan. 4900 m. Aug.–Sep. **Pl. 1.**

4 **Adonis aestivalis**   See F.H. p. 12. **Pl. 2.**

5 **Anemone biflora**   See F.H. p. 12. **Pl. 2.**

6 **Anemone rivularis**   See F.H. p. 13, p. 446. **Pl. 2.**

7 **Anemone rupestris** Wallich ex Hook.f. & Thoms.
Perennial with stout fibrous rootstock, and stem 3–15 cm. Basal leaves long-stalked, with 3 lobed stalked segments; involucral leaves 3, whorled, stalkless. Flowers erect, 7–14 mm across, white with a bluish or reddish tinge. Achenes hooked, hairless.
    C. Nepal to S.E. Tibet. 3500–4600 m. Jun.–Jul. **Pl. 1.**

8 **Anemone vitifolia**   See F.H. p. 12, p. 446. **Pl. 2.**

9 **Clematis acuminata** DC.
Slender branched climber to 4 m, hairless except the flowers. Leaves divided into 3 ovate-lanceolate long-pointed, entire or distantly toothed, shining leaflets; terminal leaflet larger. Flowers creamy-white, nodding, 2.5–3.5 cm across, in few-flowered branched clusters. Sepals oblong-lanceolate, downy inside. Filaments hairy. Achenes silky, with long feathery styles.
    Uttar Pradesh to Bhutan. 1000–2400 m. Nov.–Feb. **Pl. 3.**

1

DILLENIACEAE

10 **Clematis alternata** Kitam. & Tamura
Slender softly hairy climber. Leaves 3–5–lobed, cordate at base, apex long-pointed, irregularly toothed, silvery-haired beneath, long-stalked. Flowers pinkish-red, *c.* 2.5 cm long, nodding, on slender axillary stalks. Sepals oblong acute, reflexed at tip, hairy. Filaments thread-like, with dilated bases.
Endemic to C. Nepal. 1800–3000 m. Jul.–Sep. **Pl. 3.**

11 **Clematis buchananiana** See F.H. p. 17. **Pl. 3.**

12 **Clematis tibetana** See F.H. p. 17. **Pl. 3.**

13 **Delphinium glaciale** Hook.f. & Thoms.
Softly hairy perennial with leafy stem to 30 cm, but often dwarfed at high altitudes. Leaves 3–lobed; lobes deeply cut into many toothed segments. Flowers mauvish-blue, inflated, 2.5–4 cm, hairy, in few-flowered clusters. Bracts leafy. Bracteoles linear or with linear lobes. Sepals orbicular. Spur to 1 cm, straight, conical. Follicles 4–5.
C. Nepal to Bhutan. 3300–5500 m. Jul.–Sep. **Pl. 1.**

14 **Delphinium nepalense** Kitam. & Tamura.
Stem 10–20 cm, flexuous, covered with reflexed white hairs. Lower leaves with 3–5 obovate-rhomboid lobes with toothed margins, hairy, long-stalked; upper leaves with stalks dilated at base. Bracts ovate-lanceolate, entire. Flowers pale or dark violet-blue, inflated, papery, solitary, clothed with pale yellow or whitish hairs, 2.5–3 cm long and broad, with conic-cylindric spur *c.* 4 mm.
E. Nepal to Bhutan. 4000–5500 m. Aug.–Oct. **Pl. 1.**

15 **Delphinium stapeliosum** Brühl ex Huth
Stem erect, branched, to 1 m, covered with stiff backward-pointing hairs. Leaves rounded-kidney-shaped, deeply cut into 3 segments, coarsely toothed, paler beneath; lateral segments unequally 2–lobed. Flowers purple or deep blue, *c.* 3 cm across, in large lax clusters. Spur awl-shaped, 1–2 cm. Follicles cylindric. Seeds winged along the edges.
W. to E. Nepal. 1200–3000 m. Jul.–Oct. **Pl. 2.**

16 **Ranunculus diffusus** See F.H. p. 10. **Pl. 5.**

17 **Ranunculus tricuspis** See F.H. p. 11. **Pl. 6.**

18 **Thalictrum foliolosum** See F.H. p. 15. **Pl. 1.**

DILLENIACEAE

19 **Dillenia pentagyna** Roxb.
Tree to 10 m with rough scarred branches, leafless when in flower. Leaves to 1 m long, oblong-lanceolate acute, strongly toothed, with 30–40 prominent paired nerves; leaf-stalk short, winged, eared at base. Flowers

2

yellow, 2.5 cm across, in umbels of 5–8 on slender stalks from warty bases. Sepals elliptic, hairless. Petals obovate. Stamens many. Fruit enclosed by enlarged fleshy sepals.
Uttar Pradesh to Bhutan. To 500 m. (India, China) Mar. Pl. 6.

MAGNOLIACEAE   Magnolia Family

20 **Magnolia campbellii** See F.H. p. 19, p. 446. Pl. 6.

21 **Michelia kisopa** See F.H. p. 19. Pl. 6.

MENISPERMACEAE

22 **Cocculus laurifolius** DC.
Straggling shrub or small tree, with smooth green twigs, and bark with raised lenticels. Leaves elliptic-lanceolate, entire, with 3 strong nerves, glossy above, paler beneath; leaf-stalk flattened on upper surface. Flowers minute, yellowish-green, in spike-like axillary clusters to 5 cm long. Inner sepals twice as long as outer. Petals 2–lobed. Fruit globose, black, shining, juicy.
Kashmir to C. Nepal. Bhutan to S.W. China. 1000–1500 m. (India, S.E. Asia) Apr.–May. Pl. 5.

BERBERIDACEAE   Barberry Family

23 **Berberis asiatica** See F.H. p. 20. Pl. 4.

24 **Berberis erythroclada** See F.H. p. 21. Pl. 4.

25 **Berberis insignis** Hook.f. & Thoms.
Evergreen undershrub to 2 m, with sparingly spiny branches. Leaves elliptic or linear-lanceolate, long-pointed, shining above and beneath, with many spiny teeth on margins; leaf-stalk red. Flowers yellow, c. 5 mm across, on short thick red flower-stalks in dense axillary clusters 2–3 cm across. Petals 2–lobed. Berry ovoid, black, bloomed.
E. Nepal to Bhutan. 2000–4000 m. Apr.–May. Pl. 4.

26 **Berberis koehniana** C.K. Schneider
Deciduous shrub to 2 m with red spiny branchlets. Leaves obovate-oblong or oblanceolate, with tapering bases; margins variably spiny, but always with spiny tip. Flowers yellow, c. 5 mm across, on slender stalks in lax branching clusters much longer than leaves. Berry ovoid, red.
Uttar Pradesh to C. Nepal. 2500–3500 m. Jun.–Aug. Pl. 5.

27 **Berberis lycium** See F.H. p. 21. Pl. 4.

28 **Berberis mucrifolia** Ahrendt
Deciduous shrub 30–60 cm of drier areas, with yellow angled finely grooved branches and long slender spines. Leaves narrowly elliptic-obovate, with spiny tip, pale beneath, turning bright red in autumn. Flowers usually solitary, rarely 2, yellow, 9 mm across, on stalks much shorter than the spines. Sepals oblong-ovate. Petals oblong-ovate, base wedge-shaped, apex entire. Berry bright red, oblong-globose.
  W. to E. Nepal. 2100–4500 m. Apr.–Jun. Pl. 4.

PAPAVERACEAE   Poppy Family

29 **Argemone mexicana** PRICKLY POPPY   See F.H. p. 25. Pl. 8.

30 **Corydalis alburyi** Ludlow
High-altitude scree-plant with long branching roots, and with erect stem 10–15 cm increasing in thickness from base to apex. Basal leaves 1–3, grey-green or sometimes purplish, with 3 leaflets deeply divided into obovate entire lobes. Stem-leaves nil, or one small bract-like beneath inflorescence. Flowers slaty-blue to whitish, with green markings, 3–4 in crowded terminal cluster. Sepals minute, toothed. Dorsal petal 22 mm long including spur 4.5 mm.
  C. Nepal endemic. 5000 m. Jun.–Jul. Pl. 7.

31 **Corydalis gerdae** Fedde
Dwarf high-altitude scree-plant, with long rootstock clothed at top with remains of old leaves. Leaves grey-green, 1–2–pinnate, long-stalked; ultimate segments with 3 oblong rounded lobes. Flowers mauve-violet marked with purple and green, 1–1.5 cm long, in dense clusters. Bracts wedge-shaped or obovate-oblong, entire or palmately lobed. Sepals toothed. Spur thick, funnel-shaped, curved at tip.
  W. Nepal to Bhutan. 4300–5500 m. Jul.–Aug. Pl. 7.

32 **Corydalis latiflora** Hook.f. & Thoms.
High-altitude scree-plant with long branching rootstock clothed with many papery scales, and with simple stem 5–10 cm. Leaves grey-green, long-stalked, 2–3–pinnate; ultimate segments linear-oblong, acute or rounded at apex. Stem-leaves 2, opposite, borne close beneath the inflorescence. Flowers *c.* 18 mm long, pale blue or mauve, with yellow or greenish tips, 3–6 in a cluster. Bracts linear. Dorsal petal broadly winged, much larger than the blunt spur.
  W. Nepal to Sikkim. 4200–5500 m. Jun.–Sep. Pl. 7.

33 **Corydalis megacalyx** Ludlow
Dwarf scree-plant 5–12 cm, with long rootstock and finely cut grey-green leaves. Leaves pinnate; leaflets deeply lobed or pinnately cut; ultimate segments minute, linear. Flowers yellow striped with dark brown, *c.* 15 mm

long, in dense cluster partly hidden by leaves. Bracts subtending flowers similar to stem-leaves but smaller and with minutely toothed bases. Sepals with irregularly toothed margins. Spur cylindric, straight.

W. to C. Nepal. 3600–4500 m. Jul. Pl. 8.

34 **Meconopsis bella** See F.H. p. 26, p. 447. Pl. 7.

35 **Meconopsis napaulensis** See F.H. p. 27. Pl. 7.

CRUCIFERAE   Mustard Family

36 **Arabidopsis himalaica** See F.H. p. 38, p. 449. Pl. 8.

37 **Arabis pterosperma** Edgew. (*A. alpina* auct. non L.)
Annual or biennial with one or several erect usually unbranched stems 20–40 cm clothed with forked hairs. Basal leaves in a rosette, oblong-obovate, short-stalked or narrowed to base; stem-leaves oblong, stalkless, more or less eared at base; all leaves toothed or entire, with stellate hairs. Flowers white, on stalks longer than calyx, in terminal clusters. Sepals ovate, hairless. Petals oblong, 5–8 mm, narrowed to base. Pod narrow-linear, compressed, erect when young.

Kashmir to Bhutan. 1200–4000 m. Apr.–Jun. Pl. 10.

38 **Arcyosperma primulifolium** See F.H. p. 39. Pl. 9.

39 **Barbarea vulgaris** R.Br.   WINTER CRESS
Biennial with erect hairless stem 15–60 cm. Lower leaves lyrate-pinnate, stalked; lateral lobes oblong, terminal lobe larger, rounded or cordate. Upper leaves obovate-oblong, entire or toothed, stem-clasping. Flowers yellow, 7–9 mm across, in crowded axillary and terminal clusters. Petals twice as long as sepals. Fruit narrowly linear, 2–3.5 cm, more or less erect, 4-angled.

Pakistan to Uttar Pradesh. 2400–3300 m. (W. Asia, Europe) May–Jun. Pl. 5.

40 **Brassica juncea** MUSTARD   See F.H. p. 43. Pl. 109.

41 **Cardamine macrophylla** See F.H. p. 41, p. 451.
The E. Himalayan form of this species is larger and more robust, with stems to 1 m or more and dark lilac flowers. The W. Himalayan form, common in Kashmir, is smaller and usually with pale mauve or white flowers. The plant shown grew in E. Nepal. Pl. 10.

42 **Cardamine violacea** See F.H. p. 41, p. 451. Pl. 10.

43 **Crambe kotschyana** See F.H. p. 34. Pl. 9.

44 **Ermania himalayensis** See F.H. p. 39, p. 450. Pl. 9.

45 **Erysimum chamaephyton** Maxim.
Dwarf stemless high-altitude perennial with branching rootstock. Leaves

in a rosette, grey-green, adpressed-hairy, linear-oblong narrowed to base, entire. Flowers rose-pink and white, in crowded central cluster. Sepals 3.5 mm long, oblong blunt, concave. Petals 6 mm long, spathulate, with long claw, apex cut-off or rounded. Pod narrow-oblong, 4–angled.

Tibet. 5200 m. Jun. Not yet recorded from our area, but since the photo was taken S.W. of Shigatse not far from the Nepal border it may perhaps occur there. Pl. 9.

46 Lignariella hobsonii See F.H. p. 42, p. 451. Pl. 10.·

47 Pegaeophyton scapiflorum See F.H. p. 37, p. 449. Pl. 8.

48 Staintoniella nepalensis Hara
Dwarf hairless high-altitude scree-plant with long branching rootstock. Leaves all basal, slightly fleshy, sparsely gland-dotted, ovate entire, apex rounded, long-stalked. Flowers pale purple veined with deeper purple, or slaty-blue, in a terminal cluster borne on leafless stem 2–8 cm. Bracts subtending inflorescence ovate-oblong. Sepals erect or spreading, elliptic, white-margined. Petals obovate, with long clawed base. Pod obovate, compressed, crowned with slender style.

W. to C. Nepal. 5000–5800 m. Jun.–Jul. Pl. 9.

49 Thlaspi cochleariforme DC.
Tufted perennial with stiffly erect hairless simple or branched stem to 30 cm. Basal leaves in rosette, oblong-ovate or rounded, long-stalked; stem-leaves oblong-ovate, base stem-clasping and with small blunt ears. Flowers c. 4 mm long, in compact terminal cluster to 1.5 cm across. Petals white veined with purple. Pod obovate, compressed, narrowed to base, with persistent style.

Pakistan to Uttar Pradesh. 1800–4300 m. May–Jul. Pl. 10.

CAPPARIDACEAE Caper Family

50 Capparis zeylanica L. (*C. horrida* L.f.)
Climbing shrub armed with stout broad-based recurved thorns. Young buds, shoots and leaves brown-woolly. Leaves ovate obovate or oblong, net-veined, apex blunt acute or fine-pointed. Flower-buds globose. Flowers 3.5–5 cm across, white or purplish, on long supra-axillary stalks. Sepals 4, free. Petals 4, oblong, hair-fringed. Stamens much longer than petals. Fruit red-brown, fleshy, bluntly 4–angled.

Kashmir to E. Nepal. To 1000 m. (India, S.E. Asia) Feb.–Mar. Pl. 14.

FLACOURTIACEAE

51 Casearia glomerata Roxb.
Shrub or small tree to 12 m. Leaves elliptic-lanceolate long-pointed,

toothed (often obscurely so), acute or blunt at base. Flowers greenish-yellow, *c.* 5 mm across, clustered in leaf-axils. Calyx deeply 4–5–lobed. Petals nil. Stamens 7–10. Staminodes yellow. Fruit ellipsoid, succulent, orange when ripe.

Himachal Pradesh to Arunachal Pradesh. 600–1700 m. (India, Burma, S. China) Mar.–Apr. Pl. 5.

POLYGALACEAE   Milkwort Family

52 **Polygala arillata**   See F.H. p. 47, p. 452. Pl. 13.

CARYOPHYLLACEAE   Pink or Carnation Family

53 **Arenaria edgeworthiana**   See F.H. p. 50. Pl. 12.

54 **Arenaria festucoides**   See F.H. p. 49. Pl. 12.

55 **Arenaria globiflora**   See F.H. p. 50. Pl. 12.

56 **Arenaria polytrichoides**   See F.H. p. 49. Pl. 11.

57 **Silene fissicalyx** Bocquet & Chater
Tufted perennial with branching rootstock and many glandular-hairy simple stems 10–25 cm. Leaves hoary, broadly elliptic-ovate, acute or blunt; lower leaves stalked, upper stalkless. Flowers nodding, solitary. Calyx inflated, globose, papery, to 1.25 cm long, pale green ribbed with purple, cut to well below middle into 5 oblong blunt hairy lobes.

C. Nepal endemic. Recorded only from Gosaikund and Mailung Khola. 4300 m. Aug.–Sep. Pl. 11.

58 **Silene gonosperma** subsp. **himalayensis**   See F.H. p. 53. Pl. 11.

59 **Silene longicarpophora** (Komarov) Bocquet
Sticky-hairy perennial scree-plant of Ladakh, with stout rootstock and several erect simple stems 10–30 cm. Leaves elliptic or spathulate, hair-fringed, 1–nerved. Flowers somewhat nodding, solitary or in clusters of 2–3. Calyx *c.* 2 cm long, inflated, hairy, with brown or violet nerves. Petals white tinged with purple. This species, together with others in the same group which also occur in Ladakh, appear to be in need of botanical revision.

Afghanistan to Kashmir. 3600–4800 m. (C. Asia) Jul.–Aug. Pl. 11.

60 **Silene nigrescens**   See F.H. p. 53. Pl. 11.

61 **Silene tenuis**   See F.H. p. 54. Pl. 13.

62 **Stellaria congestiflora** Hara
Mat-forming perennial of drier areas, with densely tufted stems to 20 cm, white-woolly in upper part. Leaves paired, narrow-lanceolate, 1–nerved, with long rigid point at apex, stalkless. Flowers white, in dense leafy

terminal clusters. Sepals 5, *c.* 5 mm long, lanceolate acute, white-margined, hairless. Petals 5, much shorter than sepals, cut almost to base into 2 linear-lanceolate lobes. Stamens 10.

W. to C. Nepal. Local endemic north of Annapurna and Dhaulagiri. 4200–4700 m. Jun.–Aug. Pl. 12.

TAMARICACEAE   Tamarisk Family

63 **Myricaria elegans** See F.H. p. 55.
This species has recently been renamed *Tamaricaria elegans* (Royle) Quaiser & Ali. Pl. 15.

GUTTIFERAE   St John's Wort Family

64 **Hypericum cordifolium** Choisy
Shrub to 120 cm, erect or hanging from steep banks, with long slender red-brown branches. Leaves oblong-lanceolate acute, shallowly cordate at base, blue-green beneath, stalkless. Flowers yellow, 3.8 cm across, in many-flowered clusters. Sepals lanceolate. Petals obliquely obovate, with prominent short sharp apical point. Stamens as long as style, 2/3 as long as petals and 1½ times as long as ovary.

C. Nepal. Common in Nepal valley and surrounding areas. 900–1900 m. Mar.–Apr. Pl. 14.

65 **Hypericum hookerianum** See F.H. p. 57. Pl. 13.

66 **Hypericum podocarpoides** See F.H. p. 56. Pl. 14.

67 **Hypericum uralum** See F.H. p. 57. Pl. 14.

THEACEAE   Tea or Camellia Family

68 **Eurya acuminata** See F.H. p. 58, p. 453. Pl. 15.

69 **Eurya cerasifolia** (D. Don) Kubuski (*E. symplocina* Blume p.p.)
Slender evergreen shrub, to 6 m but usually smaller, with furrowed branches silky at tips. Leaves oblong-elliptic, entire or toothed, downy beneath on midrib. Flowers white, *c.* 5 mm across, in crowded short-stalked axillary clusters. Petals 5. Fruit a fleshy berry.

C. Nepal to S.W. China. 900–2300 m. Apr.–May and Sep.–Nov. Pl. 15.

MALVACEAE   Mallow Family

70 **Abelmoschus manihot** See F.H. p. 62. Pl. 13.

71   **Hibiscus trionum** L.
Annual weed of fields, with downy or hairy stem to 60 cm. Lower leaves orbicular; upper leaves deeply divided into oblong blunt coarsely toothed lobes, the central lobe longest. Flowers to 4 cm across, pale yellow with purple centre, solitary, axillary. Bracteoles numerous, linear. Calyx campanulate, inflated, with bristly-hairy nerves and triangular acute lobes. Stamens united into a tube bearing numerous stalkless anthers. Capsule oblong blunt.
Afghanistan to Kashmir. To 2000 m. (India, warmer parts of Eurasia) Jul.–Sep. **Pl. 14.**

BOMBACACEAE

72   **Bombax ceiba** Silk-Cotton Tree, 'Simal'   See F.H. p. 63. **Pl. 14, Pl. 15.**

STERCULIACEAE

73   **Abroma angustifolia** (L.) L.f.
Shrub to 2.5 m with downy branches. Leaves ovate-oblong long-pointed, base cordate and 3–7–nerved, margins unevenly toothed. Flowers maroon, to 5 cm across, nodding, in few-flowered axillary clusters. Sepals lanceolate, fused at base. Petals 5, soon falling, concave below, prolonged above into a spoon-shaped blade. Capsule papery, 5–winged, cut-off at apex.
Pakistan to Sikkim. Indigenous, also cultivated for fibre and medicine. To 1100 m. Jun.–Sep. **Pl. 16.**

74   **Reevesia** sp.
The genus is distinguished by the long staminal column protruding from the flower which bears a globose head of anthers. Dried specimens of the plant illustrated do not match the descriptions either of *Reevesia wallichii* or *Reevesia pubescens*. They were collected from a single tree 15 m tall growing above the bridge over the Arun near Num.
E. Nepal. 1200 m. May. **Pl. 17.**

TILIACEAE   Lime Family

75   **Grewia optiva** J.R. Drumm. ex Burret (*G. oppositifolia* Buch.–Ham. ex D. Don)
Deciduous shrub or small tree to 12 m, planted in villages and lopped for fodder. Leaves alternate, ovate long-pointed, with small close-set blunt teeth, 3–nerved from base, stalked. Flowers cream or white, in 1–8–

flowered clusters on hairy leaf-opposed stems. Sepals 18 mm, linear-oblong, 3–ribbed, red within. Petals shorter than sepals, linear, with distinct claw. Fruit 1–4–lobed, black when ripe.
Pakistan to Sikkim. To 1800 m. Apr.–May. Pl. 16.

MALPIGHIACEAE

76 Hiptage benghalensis (L.) Kurz (*H. madablota* Gaertn.)
Large evergreen climber. Leaves oblong or ovate-lanceolate, long-pointed, entire, hairless. Flowers to 2.5 cm across, in axillary clusters to 15 cm long. Calyx-lobes oblong blunt. Petals white, fringed, reflexed, the fifth petal yellow at base. Stamens 10, curved, the 2 lowest twice as long as the others. Fruit of 1–3 globose winged nuts.
Pakistan to Bhutan. To 1100 m. (India, China, S.E. Asia) Mar.–Apr. Pl. 16.

GERANIACEAE   Geranium Family

77 Geranium lambertii   See F.H. p. 67. **Pl. 17.**
78 Geranium pratense   Meadow Cranesbill   See F.H. p. 67. **Pl. 16.**
79 Geranium procurrens   See F.H. p. 66. **Pl. 17.**

OXALIDACEAE

80 Oxalis acetosella   Wood-Sorrel   See F.H. p. 68. **Pl. 17.**
81 Oxalis latifolia   See F.H. p. 68. **Pl. 17.**

RUTACEAE   Citrus Family

82 Micromelum integerrimum (Buch.–Ham.) Wight & Arn. ex M. Roem.
(*M. pubescens* auct. non Blume)
Small evergreen tree. Leaves pinnate, densely gland-dotted, aromatic when crushed; leaflets obliquely ovate-lanceolate, margins undulate or obscurely rounded-toothed. Flowers white, fragrant, 8–12 mm across, in branched spreading flat-topped clusters. Calyx small, with triangular lobes. Petals narrow-oblong, downy. Stamens 10, alternately long and short. Berry oblong, orange-yellow when ripe.
C. Nepal to Bhutan. To 1200 m. (India, Burma) Dec.–Feb. Pl. 19.

MELIACEAE   Mahogany Family

83 **Toona serrata** (Royle) M. Roem. (*Cedrela s.* Royle) HILL 'TOON'
Deciduous tree to 15 m. Leaves large, pinnate; leaflets elliptic-oblong long-pointed, irregularly toothed, base oblique; leaf-stalk and rhachis dark red. Flowers pinkish-white, 5 mm long, in branched drooping clusters to 1.5 m long. Calyx small, shortly lobed. Petals elliptic-oblong blunt. Fruit an ovoid 5-valved capsule.
Pakistan to C. Nepal. 1500–2300 m. Jun.–Jul. **Pl. 19.**

AQUIFOLIACEAE   Holly Family

84 **Ilex excelsa** (Wallich) Hook.f.
Medium-sized evergreen tree. Leaves ovate-elliptic long-pointed, entire, with arching lateral veins; leaf-stalk channelled above. Flowers greenish-white, 3.5–4 mm across, in axillary umbels. Calyx with 4–5 ovate blunt hair-fringed lobes. Petals 4–5, ovate-orbicular, fused at base. Fruit globose, bright red.
Himachal Pradesh to S.W. China. 600–2100 m. May–Jun. **Pl. 18.**

CELASTRACEAE   Spindle-Tree Family

85 **Euonymus fimbriatus**   See F.H. p. 76. **Pl. 18.**

86 **Euonymus grandiflorus** Wallich
Deciduous shrub or small tree. Leaves obovate-elliptic, finely toothed, dark glossy green above, stalked. Flowers *c.* 2.5 cm across, in erect axillary 3–7–flowered clusters. Petals 4, orbicular, pale greenish-yellow. Disk bright green, flat, *c.* 6 mm across. Capsule pink when ripe, deeply 4–lobed but not winged.
Uttar Pradesh to S.W. China. 1500–1800 m. May–Jun. **Pl. 18.**

87 **Euonymus pendulus** Wallich
Small evergreen tree of forest understory, or sometimes epiphytic. Branchlets slender, grey, pendulous. Leaves oblong-lanceolate, narrowed at both ends, sharply toothed, leathery, glossy above, paler beneath. Flowers in branching clusters, at tips or bases of shoots. Petals white, 6 mm, ovate-oblong, fringed, spreading. Capsule winged.
Pakistan to Arunachal Pradesh. 1800–2600 m. Apr.–Jun. **Pl. 18.**

RHAMNACEAE   Buckthorn Family

88 **Zizyphus mauritiana** Lam. (*Z. jujuba* (L.) Gaertn.)
More or less evergreen shrub, to 4 m but usually smaller, with spreading

11

drooping softly hairy branches armed with curved prickles. Leaves ellip-tic-ovate or orbicular, entire or regularly toothed, unequal at base, dark green and with 3 prominent veins above, pale and woolly-hairy beneath. Flowers *c.* 3 mm across, greenish-yellow, in axillary clusters. Petals spathulate, concave, reflexed. Fruit globose, fleshy, yellow becoming orange-red. Bark, leaves and fruit are used medicinally.

Pakistan to Bhutan. 200–1200 m. (Tropical Asia, Australia) May–Sep. Pl. 19.

VITACEAE   Vine Family

89 **Cissus javana** DC. (*Vitis discolor* (Blume) Dalz.)
Weak climber with woody base, red hairless stems, and forked tendrils. Leaves ovate-lanceolate acute, base often cordate, margins fine-toothed or bristly, usually mottled with white above, purple and with arching nerves beneath, stalked. Flowers minute, yellowish, in small compound umbels on bright red leaf-opposed stalks. Berry black to reddish-purple.

C. Nepal to Sikkim. S.W. China. To 1200 m. (S.E. Asia) Jul.–Aug. Pl. 19.

ANACARDIACEAE   Mango or Cashew Family

90 **Cotinus coggygria**   Wig Tree, Smoke Tree   See F.H. p. 83, p. 458. Pl. 8.

91 **Dobinea vulgaris** Buch.–Ham. ex D. Don
Lax shrub to 4 m, much more prominent when in fruit in late autumn than when in flower. Leaves opposite, elliptic-lanceolate long-pointed, regu-larly toothed, stalked. Flowers tiny, *c.* 1 mm across, pale yellow, in long drooping widely branched clusters. Fruit a small compressed capsule subtended by conspicuous yellowish-white or rose-coloured veined papery obcordate-orbicular bracts.

C. Nepal to Bhutan. 1500–2300 m. Aug.–Oct. Pl. 20.

92 **Lannea coromandelica** (Houtt.) Merr. (*Odina wodier* Roxb.)
Tree to 20 m with thick trunk and few branches, leafless when in flower. Leaves pinnate, crowded towards ends of branches; leaflets ovate-oblong long-pointed, entire, base oblique. Flowers pale yellow, *c.* 5 mm across, unisexual, in long slender inflorescences crowded towards tips of leafless branches. Male inflorescence compound, female simple. Calyx tubular, with 4 rounded hair-fringed lobes. Petals 4, ovate-oblong, spreading. Fruit oblong, fleshy, red when ripe. Bark and leaves are used medicinally.

Pakistan to Bhutan. To 1400 m. (China, Burma, S.E. Asia) Mar.–Apr. Pl. 6.

93 **Rhus wallichii** See F.H. p. 84, p. 458. Pl. 20.

CORIARIACEAE   Coriaria Family

94 **Coriaria napalensis** See F.H. p. 85, p. 458. Pl. 20.

MORINGACEAE

95 **Moringa oleifera** Lam. (*M. pterygosperma* Gaertn.) HORSERADISH TREE
Small deciduous tree with corky bark. Leaves 2–3–pinnate; leaflets entire, paler beneath; lateral leaflets elliptic, terminal leaflet slightly larger, obovate. Flowers white, *c.* 2.5 cm across, fragrant, in large branched axillary downy clusters. Calyx-lobes linear-lanceolate, reflexed, downy. Petals 5, white with yellow dots at base, linear-spathulate; upper pair smaller, lateral pair ascending, anterior petal larger. Filaments hairy at base. Capsule pendulous, to 35 cm long, beaked, 9–ribbed. Leaves, flowers and young fruit are eaten as vegetables. The young roots when scraped furnish a good substitute for horseradish.
  Indigenous to W. Himalaya. Widely cultivated throughout foothills. To 1100 m. Feb.–Apr. Pl. 20.

LEGUMINOSAE   Pea Family

Subfamily CAESALPINIOIDEAE

96 **Bauhinia variegata** See F.H. p. 88. Pl. 21.

97 **Cassia floribunda** Cav.
Shrub to 2 m or sometimes an annual, gregarious as a weed on wasteland. Leaves pinnate, with grooved leaf-stalk; leaflets ovate-elliptic long-pointed, veins prominent above, with conical gland on rhachis at base. Flowers bright yellow, *c.* 1.5 cm across, in axillary and terminal clusters. Sepals oblong-ovate, papery. Petals ovate-orbicular. Three stamens much longer than the others, and with deeply curved filaments. Pod cylindric, transversely partitioned.
  Probably native of tropical America. Naturalized in Himalaya to 2200 m. Jun.–Sep. Pl. 21.

98 **Caesalpinia cucullata** Roxb. (*Mezoneurum c.* (Roxb.) Wight & Arn.)
Rambling shrub armed with hooked prickles. Leaves 2–pinnate; leaflets ovate long-pointed, blue-green beneath. Flowers in simple or forked axillary and terminal clusters. Calyx yellow, 9 mm, very oblique, deeply lobed; lowest lobe larger and forming a hood. Petals 5, yellow streaked with red; uppermost petal folded, 2–lobed. Stamens long-exserted, filaments and anthers red. Pod pinkish, large flat oblong, 1–seeded, with broad papery wing down one side.

13

LEGUMINOSAE

W. Nepal to Bhutan. To 1800 m. (India, China, S.E. Asia) Oct.–Feb.
Pl. 21.

Subfamily PAPILIONOIDEAE

99   Apios carnea   See F.H. p. 102, p. 463. Pl. 24.

100 Astragalus chlorostachys   See F.H. p. 105. Pl. 22.

101 Astragalus donianus   See F.H. p. 104. Pl. 22.

102 Astragalus frigidus   See F.H. p. 105. Pl. 22.

103 Astragalus melanostachys Benth. ex Bunge
Perennial of drier areas with stout branching rootstock and spreading or
procumbent stems to 50 cm. Leaves pinnate; leaflets blue-green, oblong,
apex blunt or notched. Stipules triangular, sharp-pointed. Flowers purple,
5–7 mm long, in elongated clusters 2.5–7.5 cm long borne on black-hairy
stalks much longer than leaves. Calyx hairy, with teeth as long as tube.
Corolla narrow; keel and wings shorter than standard. Pod oblong,
2–seeded, covered with black silky hairs.
        Afghanistan to W. Nepal. 3300–5100 m. (Tibet, C. Asia) Jul.–Aug. Pl.
23.

104 Astragalus subuliformis DC. (A. subulatus M. Bieb.)
Perennial of drier areas with slender forked adpressed-white-hairy stems
to 20 cm. Leaves pinnate; leaflets distant, linear-awl-shaped, adpressed-
hairy above and beneath. Stipules minute, linear. Flowers pale yellow, c.
2.5 cm, in few-flowered clusters borne on hairy stalks longer than leaves.
Bracts persistent, minute, lanceolate. Calyx hairy, cylindric, with short
awl-shaped teeth. Corolla narrow. Pod hairy, cylindric, beaked, straight
or slightly curved.
        Afghanistan to Kashmir. 2700–3300 m. (W. Asia) Jun.–Aug. Pl. 22.

105 Butea monosperma (Lam.) Kuntze (B. frondosa Koen. ex Roxb.)
Deciduous tree to 15 m, flowering when almost leafless. Leaves trifoliate,
with long leaf-stalk swollen at base; terminal leaflet broadly obovate from
wedge-shaped base, or rhomboid; lateral leaflets smaller, obliquely ovate;
all silky beneath. Flowers bright red tinged with orange, 4–5 cm long, in
rigid brown-velvety axillary and terminal clusters. Calyx broad-cam-
panulate, grey-silky inside, with short triangular teeth. Corolla silky-
haired outside; standard ovate acute, recurved; wings sickle-shaped; keel
curved and beaked. Pod oblong, with ribbed margins and pointed tip,
densely covered with silvery hairs. Bark, leaves and seed are used
medicinally.
        Pakistan to E. Nepal. To 1200 m. (India, S.E. Asia) Feb.–Apr. Pl. 23.

106 Campylotropis speciosa   See F.H. p. 92, p. 461. Pl. 23.

107 **Caragana jubata** See F.H. p. 97. **Pl. 23.**

108 **Caragana sukiensis** See F.H. p. 97. **Pl. 24.**

109 **Cochlianthus gracilis** Benth.
Slender climber to 2 m. Leaves trifoliate; leaflets thin, with a few adpressed hairs on both surfaces, grey beneath; terminal leaflet ovate-rhomboid long-pointed, lateral leaflets unequal at base. Flowers mauve or pink, to 2 cm, in short dense clusters on slender drooping stalks. Calyx silky-haired, campanulate, with lower teeth longer than the fused upper ones. Standard of corolla broad; keel narrow, curved, rounded at end. Pod linear, with curved apex, densely brown-hairy.
C. Nepal to S.W. China. 1800–2000 m. Jul.–Sep. **Pl. 24.**

110 **Crotalaria cytisoides** See F.H. p. 92, p. 459. **Pl. 24.**

111 **Crotalaria tetragona** Andrews
Weak shrub to 2 m with 4–angled thinly silky branches. Leaves linear-lanceolate long-pointed, entire, silky-haired beneath, short-stalked. Flowers bright yellow, *c.* 2 cm long, in long lax terminal and axillary clusters. Bracts minute, bristle-like. Calyx velvety, usually reddish-brown, with linear-lanceolate long-pointed teeth. Standard orbicular; keel much incurved and sharply beaked. Pod linear-oblong, dark velvety brown.
Uttar Pradesh to Bhutan. To 1700 m. Aug.–Nov. **Pl. 24.**

112 **Derris microptera** Benth.
Large climber, showy when in flower. Leaves pinnate; leaflets obovate-oblong long-pointed, thin and hairless. Flowers red or mauvish-pink, *c.* 1.3 cm, in long flexuous clusters. Calyx broadly campanulate, with short triangular lobes. Standard erect. Pod hairless, oblong flat, 1–2–seeded, with very narrow wing.
E. Nepal to Sikkim. 600–1500 m. Apr.–May. **Pl. 25.**

113 **Desmodium confertum** DC.
Shrub to 1½ m, with branchlets finely downy when young. Leaves trifoliate; leaflets obovate, silky-haired and with raised veins beneath. Flowers pinkish-mauve, to 12 mm, in dense axillary and terminal clusters. Bracts papery, hair-fringed. Calyx top-shaped, shortly toothed. Pod 3–4–jointed, densely hairy.
C. Nepal to Arunachal Pradesh. 300–2000 m. Jul.–Nov. **Pl. 26.**

114 **Desmodium multiflorum** See F.H. p. 93. **Pl. 26.**

115 **Hedysarum kumaonense** See F.H. p. 108.
The name in F.H. is misprinted. The above spelling is correct. **Pl. 26.**

116 **Hedysarum manaslense** (Kitam.) Ohashi
Perennial with robust stem 30–150 cm, densely white-woolly at first. Leaves pinnate; leaflets narrowly ovate-elliptic, fine-pointed at apex, entire, woolly beneath. Stipules leaf-opposed, papery, fused at base. Flowers purple-pink, 1.5–2 cm long, in axillary long-stalked clusters.

15

Calyx downy, 4–lobed above the middle; upper lobes triangular, 2–toothed at apex; lower lobes longer and narrower. Standard obovate, shorter than keel; wings longer, narrow-elliptic blunt, eared at base. Pod ascending, 2–4–jointed, with seams slightly margined.

C. Nepal endemic. 3000–4200 m. Jun.–Aug. Pl. 25.

117 **Indigofera exilis** Grierson & Long
Shrub to 1½ m with grooved silky-haired branchlets. Leaves pinnate; leaflets elliptic-oblong blunt, apex fine-pointed and sometimes notched, adpressed-hairy above and beneath. Flowers purple-pink, *c*. 1 cm long, in long slender clusters. Calyx brown-hairy, obliquely triangular-toothed. Pod straight, swollen, hairless.

C. to E. Nepal. 1300–2100 m. May–Aug. Pl. 25.

118 **Indigofera pulchella** See F.H. p. 99. Pl. 25.

119 **Lathyrus laevigatus** See F.H. p. 100, p. 462. Pl. 21.

120 **Lotus corniculatus** COMMON BIRD'S-FOOT TREFOIL See F.H. p. 109. Pl. 27.

121 **Mucuna macrocarpa** Wallich
Large woody climber with bark exuding sticky gum when cut. Leaves trifoliate; leaflets elliptic-ovate, abruptly long-pointed, hairy above and beneath when young, with prominent arching nerves beneath; lateral leaflets oblique. Flowers to 7 cm long, in dense pendulous clusters from old wood. Calyx humped, with short unequal lobes, covered with fine sharp bristles. Corolla bristly; standard broad, pale yellow; wings fleshy, purple; keel pale yellow, abruptly turned up at tip. Pod very large, to 50 cm, channelled, velvety at first. This plant should be handled with care, because the sharp deciduous bristles can penetrate the skin and cause considerable irritation.

C. Nepal to Bhutan. 1200–1800 m. (India, China, Japan, S.E. Asia) Mar.–Apr. Pl. 26.

122 **Oxytropis arenae-ripariae** Vass.
Tufted stemless perennial of riverside gravel, with woody rootstock. Leaves pinnate; leaflets oblong acute, sparsely white-hairy beneath. Flowers bluish-mauve, *c*. 8 mm, in short dense clusters borne on spreading or decumbent silky-haired stalks which are longer than the leaves. Calyx campanulate, with linear teeth, densely covered with black hairs. Pod unknown.

W. to E. Nepal. 4500–4700 m. Jun.–Jul. Pl. 28.

123 **Oxytropis cachemiriana** See F.H. p. 106.
Despite the very different appearance of the plants shown in 123A and 123B both are the same species. 123A was growing in comparatively wet country in Kashmir on the Sonamarg side of the Zoji La. 123B grew in very dry country in Ladakh beyond Kargil on the road to Leh. Pl. 28.

124 **Oxytropis humifusa** Karelin & Kir.
Tufted stemless perennial of drier areas, with strongly branching rootstock. Leaves pinnate, grey-green; leaflets elliptic-oblong acute, densely white-hairy above and beneath. Flowers mauve, 5-7 mm, in short dense white-silky clusters 1-2 cm across borne on decumbent stalks much longer than the leaves. Bracts linear-awl-shaped. Calyx campanulate, with linear teeth, densely hairy. Standard obcordate, notched at apex; keel short, beaked. Pod ovoid-oblong, hairy.
Pakistan to Himachal Pradesh. 3000-5200 m. Jun.-Aug. Pl. 27.

125 **Oxytropis microphylla** See F.H. p. 107. Pl. 28.

126 **Oxytropis mollis** See F.H. p. 107. Pl. 27.

127 **Thermopsis inflata** See F.H. p. 96. Pl. 27.

128 **Thermopsis lanceolata** R.Br. ex Aiton
Perennial of Tibetan borderlands, with stout rootstock and erect silky-haired stems to 30 cm. Leaves trifoliate; leaflets oblong blunt, covered with adpressed hairs. Stipules ovate-lanceolate. Flowers yellow, 2 cm long, in loose terminal cluster. Bracts oblong-ovate long-pointed, adpressed-hairy above and beneath. Calyx hairy, the lanceolate lower teeth about as long as the tube. Limb of standard rounded, deeply notched; wings linear-oblong, about equalling standard; keel-petals coherent only in upper part. Pod oblong-linear, flattened, hairy, the style persisting as a beak.
W. to C. Nepal. 3600-4300 m. (Tibet, C. & N. Asia) Jun. Pl. 27.

129 **Trigonella emodi** See F.H. p. 96, p. 461. Pl. 28.

ROSACEAE   Rose Family

130 **Cotoneaster acuminatus** See F.H. p. 117. Pl. 30.

131 **Cotoneaster frigidus** See F.H. p. 119. Pl. 30.

132 **Cotoneaster integrifolius** (Roxb.) Klotz (*C. thymifolius* Baker)
Small spreading or decumbent intricately branched shrub. Leaves obovate-oblong, apex rounded, margins incurved, dark glossy green above, woolly beneath. Flowers pink in bud, white later, *c.* 8 mm across, usually solitary. Calyx·hairy. Fruit red, globose. There are ˙a number of other small decumbent *Cotoneaster* species in our area, all very difficult to distinguish. In Nepal this is the commonest low-altitude species. Very similar to *Cotoneaster microphyllus* (See F.H. p. 118), but leaves narrower, broadening upwards and with a blunt or rounded apex. (*C. microphyllus*, leaves ovate-elliptic acute).
Himachal Pradesh to S.W. China. 1800-3500 m. May-Jun. Pl. 29.

133 **Cotoneaster roseus** See F.H. p. 117. Pl. 30.

17

ROSACEAE

134   **Eriobotrya elliptica** Lindley
Small evergreen tree, indigenous but also often planted around fields.
Leaves obovate or oblong-lanceolate, entire or coarsely toothed, hairless
beneath, leathery, stalked. Flowers white, 1.5 cm across, in compact
branched clusters densely clothed in rusty wool. Calyx densely hairy, with
triangular lobes. Petals orbicular. Fruit a succulent obovoid berry.
   Uttar Pradesh to E. Nepal. 300–2100 m. Mar.–May, sometimes Oct.
Pl. 30.

135   **Maddenia himalaica** Hook.f. & Thoms.
Deciduous shrub to 6 m, with red-brown branchlets. Leaves ovate or
obovate-lanceolate, long-pointed, base rounded or cordate, margins
fringed with glandular hairs, woolly or hairless beneath. Stipules large,
linear-lanceolate long-pointed, glandular-toothed. Flowers white, 5–7
mm, in dense terminal clusters 2.5–7.5 cm long. Calyx top-shaped, with
blunt lobes. Petals minute, linear-oblong. Stamens very prominent, 20–30.
Fruit broadly ovoid, fleshy.
   E. Nepal to S.E. Tibet. 2400–3000 m. Apr.–May. **Pl. 29.**

136   **Photinia integrifolia**   See F.H. p. 120, p. 466. **Pl. 29.**

137   **Potentilla atrosanguinea**   See F.H. p. 127.
This is a very widespread and variable species. The form shown here with
dark red flowers is from Nepal, where it is not common. For an orange-
flowered form see F.H. **Pl. 37.** The much more common yellow-flowered
form should now be called var. *argyrophylla.* See Long, *Not. Roy. Bot.
Gard. Edin.*, 1979, Vol. 37, No. 2. **Pl. 29.**

138   **Potentilla plurijuga** Hand.–Mazz.
Perennial of drier areas with short rootstock and several spreading stems
to 30 cm. Leaves pinnate, softly white-woolly beneath; leaflets pinnately
cut into linear-lanceolate segments. Flowers yellow, 8–10 mm across,
solitary or few in flat-topped cluster. Calyx silky, with ovate acute lobes.
Petals orbicular-ovate, with rounded notched apices. Achenes hairless.
This species is close to *Potentilla multifida* of N. Asia & Europe, and some
botanists seem to doubt whether it merits specific status.
   Uttar Pradesh to S.W. China. 3000–4500 m. (Tibet) Jun.–Aug. **Pl. 29.**

139   **Prunus bokhariensis** Royle ex C. Schneider (*P. communis* Hudson var.
insititia Hook.f.)
Shrub to 5 m with slender blackish hairless branchlets. Leaves variable,
obovate or ovate-lanceolate, blunt acute or with fine-pointed tip, toothed,
nerves hairy beneath. Flowers appearing when shrub is still leafless,
white, on slender solitary or paired stalks. Calyx-tube short obconic.
Petals *c.* 8 mm long. Fruit a drooping globose or ovoid yellow hairless
edible plum. *The Flora of British India* quotes Dr T. Thomson as saying
'truly wild'. It is widely naturalized on hills surrounding the Kashmir

valley, but appears to be confined to localities which are not very far from areas where it is cultivated.

W. Himalaya. Cultivated to 2100 m. Apr. Pl. 32.

140 **Prunus napaulensis** See F.H. p. 116. Pl. 30.

141 **Prunus prostrata** Labill.
Spreading or procumbent deciduous shrub to 2 m, common on rocky slopes around Kashmir valley. Twigs slender, densely velvety-hairy when young. Leaves elliptic or ovate-oblong, toothed, dark green and hairless above; easily distinguished from the other dwarf pink-flowered cherry of the W. Himalaya, *Prunus jacquemontii* (See F.H. p. 115), by its leaves densely white-woolly beneath. Stipules glandular. Flowers opening while shrub is still leafless, pink, 1–3 together, almost stalkless. Calyx-tube cylindric, downy, lobes woolly-haired within. Petals rounded, *c.* 4 mm. Fruit bright red, ovoid or globose.

Afghanistan to Himachal Pradesh. 1500–3000 m. Apr. Pl. 31, Pl. 32.

142 **Pyracantha crenulata** See F.H. p. 120, Pl. 34 Pl. 32.

143 **Rubus calycinus** See F.H. p. 110, p. 464. Pl. 32.

144 **Rubus foliolosus** See F.H. p. 111. Pl. 31.

145 **Rubus hypargyrus** See F.H. p. 112. Pl. 31.

146 **Rubus nepalensis** See F.H. p. 110, p. 464. Pl. 32.

147 **Rubus niveus** Thunb. (*R. lasiocarpus* Smith)
Rambling prickly shrub to 3 m with waxy purple bloom on twigs. Leaves with channelled leaf-stalks and small linear stipules; leaflets usually 5–7, lanceolate, toothed, nerves impressed above, straight and parallel beneath, woolly-haired; terminal leaflet broadly ovate, often lobed. Flowers pink, to 1.6 cm across, in axillary and terminal clusters. Flower-stalks woolly-haired. Bracts linear. Calyx-tube short, silky-haired, with long-pointed lobes exceeding petals. Fruit black. Resembling *Rubus foliolosus*, but flowers larger and calyx silky (*R. foliolosus* flowers to 1 cm, calyx woolly-haired).

Afghanistan to Bhutan. 1800–2900 m. (India, S.E. Asia) Apr.–Jun. Pl. 31.

148 **Rubus splendidissimus** Hara (*R. andersonii* Hook.f.)
Rambling unarmed shrub to 3 m, with spreading purple gland-tipped hairs on branchlets and leaf-stalks. Leaves digitately trifoliate; central leaflet elliptic-oblong long-pointed, toothed, silvery-woolly beneath and with many prominent parallel nerves; 2 lateral leaflets similar but oblique. Flowers white, *c.* 1.5 cm across, in broad branched glandular-hairy terminal clusters. Calyx-lobes triangular-lanceolate long-pointed, glandular-hairy. Berry red.

C. Nepal to Bhutan. 2400–3000 m. Aug.–Sep. Pl. 31.

149 **Sorbus cuspidata** See F.H. p. 121, Pl. 35. **Pl. 34.**

150 **Sorbus foliolosa** See F.H. p. 122. **Pl. 33.**

151 **Sorbus insignis** (Hook.f.) Hedlund (*Pyrus i.* Hook.f.)
Tree to 10 m. Young shoots clothed with silky brown hairs, old shoots hairless. Leaves pinnate, leathery; leaflets linear-oblong, unequal at base, obscurely rounded-toothed, margins incurved, blue-green beneath. Flowers appearing with the young leaves, white, *c.* 6 mm across, in red-woolly flat-topped clusters borne on stout stalks covered with oblong white lenticels. Calyx-teeth triangular. Petals orbicular. Fruit red, globose or broadly ovoid.
E. Nepal to Sikkim. 2800–3000 m. (Burma) May–Jun. **Pl. 33.**

152 **Sorbus lanata** See F.H. p. 122, p. 465. **Pl. 34.**

153 **Sorbus microphylla** See F.H. p. 122, Pl. 36. **Pl. 33.**

154 **Sorbus ursina** See F.H. p. 122. **Pl. 33.**

155 **Sorbus wallichii** (Hook.f.) Yü (*Pyrus w.* Hook.f.)
Deciduous tree to 12 m, but often smaller and sometimes epiphytic. Young twigs densely white-woolly. Leaves pinnate; leaflets leathery, linear-oblong, acute or with fine-pointed tip, base oblique, obscurely toothed, blue-green beneath and with midrib very prominent. Flowers white, 6–8 mm across, in flat-topped compound woolly clusters to 7.5 cm across. Petals broadly obovate. Fruit ovoid, crowned with persistent calyx.
C. Nepal to S.W. China. 2400–3100 m. (Burma) Apr.–May. **Pl. 33.**

SAXIFRAGACEAE Saxifrage Family

156 **Astilbe rivularis** See F.H. p. 129, p. 466. **Pl. 34.**

157 **Bergenia purpurascens** See F.H. p. 130. **Pl. 37.**

158 **Chrysosplenium forrestii** See F.H. p. 131. **Pl. 38.**

159 **Saxifraga aristulata** Hook.f. & Thoms.
Tuft-forming perennial with erect glandular-hairy stems 3–5 cm. Basal leaves linear-spathulate. Stem-leaves narrow-linear, stalkless, hairless, with recurved margins and deciduous terminal awn. Flowers yellow, solitary. Sepals ovate, 2.5 mm long, hairless. Petals obovate, 6–8 mm.
Uttar Pradesh to Sikkim. 4200–5600 m. Jul.–Sep. **Pl. 38.**

160 **Saxifraga brachypoda** D. Don
Tufted perennial with numerous simple unbranched stems 5–20 cm, bearing solitary yellow flowers. Leaves densely overlapping, rigid, ascending or reflexed, linear-lanceolate, more or less stem-clasping at base, apex spine-tipped, margins often shortly spiny. Sepals ovate acute,

slightly downy, with spiny margins. Petals *c.* 7 mm, much exceeding sepals.

Uttar Pradesh to S.W. China. 3600–4800 m. (Burma) Jul.–Aug. Pl. 38.

161 **Saxifraga caveana** W.W. Smith (*S. diapensia* Harry Smith)
Dwarf tufted cushion-forming perennial with many glandular-hairy stems 2–5 cm. Leaves leathery, broadly ovate-lanceolate blunt, sparsely hairy, stalked; upper leaves stalkless, linear-lanceolate, hairless. Sepals ovate, margins black-glandular. Petals yellow, *c.* 10 mm, oblong-ovate.

C. Nepal to S.W. China. 4200–5000 m. Jul.–Aug. Pl. 36, Pl. 37.

162 **Saxifraga hemisphaerica** Hook.f. & Thoms.
High-altitude cushion-forming perennial with densely crowded branching stems. Leaves in dense silvery rounded rosettes 3–5 mm across; lower leaves oblong entire; upper leaves shorter, with broad white papery fringed margins. Flowers yellow, almost stalkless. Sepals ovate blunt. Petals soon falling.

Uttar Pradesh to Bhutan. 4500–5100 m. (Tibet) Jun.–Aug. Pl. 37.

163 **Saxifraga hirculoides** See F.H. p. 132. Pl. 36.

164 **Saxifraga hypostoma** Harry Smith
Perennial of drier areas forming cushions to 25 cm across. Stems densely covered with whorled oblong leaves with whitish apices. Flowers white, stalkless. Close to *Saxifraga pulvinaria* (See F.H. p. 135), but distinguished by its leaves with lower part of blade with entire hairless margins, upper part with long fringed reflexed whitish margins. Sepals ovate, *c.* 1.5 mm long and broad, margins with sparse glandular hairs. Petals obtriangular, rounded, 4 mm long and broad. Anthers red.

W. to C. Nepal. 4300–5200 m. May–Sep. Pl. 37.

165 **Saxifraga melanocentra** See F.H. p. 134. Pl. 36.

166 **Saxifraga nutans** Hook.f. & Thoms.
Stem erect, leafy, densely glandular-hairy, 12.5–25 cm. Basal leaves oblong blunt, margins gland-fringed, stalked; stem-leaves broadly oblong, stalkless, not overlapping. Flowers yellow, nodding, short-stalked, 6–12 in a cluster. Calyx dark red, densely glandular-hairy. Petals spathulate, *c.* 8 mm.

C. Nepal to S.W. China. 4200–4900 m. Jul.–Sep. Pl. 36.

167 **Saxifraga pilifera** Hook.f. & Thoms.
Stem erect, leafy, downy, 2.5–7.5 cm. Stolons red, thread-like, spreading from basal leaf-rosette and terminating in rooting buds. Leaves hairy; basal leaves oblong; stem-leaves linear-oblong. Flowers in clusters of 1–4. Sepals oblong acute, erect, 2 mm long. Petals dull brownish-red, oblong acute, 2.5 mm long, 3–nerved.

E. Nepal to Bhutan. 4800 m. Jun.–Jul. Pl. 36.

21

168 **Saxifraga punctulata** Engl.
Densely tufted plant with glandular-hairy stem 2–5 cm. Basal leaves in a
rosette, fleshy, broadly spathulate, margins glandular-hairy; stem-leaves
few, lanceolate blunt, stalkless. Flowers very pale yellow, purple-dotted
within. Sepals ovate acute, 2.5–3 mm long, margins densely glandular-
hairy. Petals oblong-elliptic, 6–7 mm long, 3–nerved.
E. Nepal to S.E. Tibet. 4900–5800 m. Sep.–Nov. Pl. 37.

HYDRANGEACEAE   Hydrangea Family

169 **Hydrangea aspera**   See F.H. p. 137. Pl. 34.

GROSSULARIACEAE   Gooseberry Family

170 **Ribes alpestre**   ASIAN GOOSEBERRY   See F.H. p. 139, p. 468. Pl. 35.

171 **Ribes glaciale**   See F.H. p. 140. Pl. 35.

172 **Ribes himalense**   See F.H. p. 139. Pl. 35.

173 **Ribes laciniatum** Hook.f. & Thoms. (*R. glaciale* var. *l.* (Hook.f. &
Thoms.) C.B. Clarke)
Unarmed shrub to 3 m, often epiphytic. Leaves on slender leaf-stalks,
triangular-ovate, cordate at base, 3–5–lobed, the lobes acute and toothed.
Flowers dark red, *c.* 5 mm across, in erect clusters to 3 cm long. Bracts
linear acute. Calyx-lobes petal-like, much larger than petals, lanceolate
acute. Fruit red, hairless.
C. Nepal to S.E. Tibet. 3000–3600 m. (Burma) Apr.–Jun. Pl. 35.

174 **Ribes orientale**   See F.H. p. 139, Pl. 44. Pl. 35.

CRASSULACEAE   Stonecrop Family

175 **Rhodiola amabilis** (H. Ohba) H. Ohba
Rhizomatous perennial forming small loose mats on damp rocks. Basal
leaves deciduous, scaly, linear-triangular acute; stem-leaves fleshy, wide-
spreading, linear-lanceolate acute, stalkless. Flowers pinkish-white, in
1–3–flowered clusters, on stalks with leafy bracts similar to stem-leaves.
Calyx fleshy, with triangular-lanceolate blunt lobes. Petals 5–10 mm,
elliptic acute.
C. to E. Nepal. 2400–3900 m. Aug.–Sep. Pl. 39.

176 **Rhodiola cretinii** (R. Hamet) H. Ohba
Small succulent mat-forming perennial with creeping stems. Leaves alternate, narrow-oblong or obovate-oblong. Flowering-stems simple, erect,
2–8 cm. Flowers unisexual, *c.* 8 mm across. Male flowers: calyx hairless,

with 5 sepals much longer than calyx-tube; corolla with 5 obovate-oblong yellow or greenish-red petals and 10 stamens. Female flowers: sepals linear blunt; petals linear-oblanceolate blunt.

E. Nepal to S.E. Tibet. 3700–4300 m. Jun.–Jul. **Pl. 39.**

177 **Rhodiola fastigiata** See F.H. p. 142. **Pl. 39.**

178 **Rosularia alpestris** See F.H. p. 141. **Pl. 38.**

179 **Sedum filipes** Hemsley
Succulent hairless perennial of mossy rocks, with erect stem to 15 cm, leafless in lower part. Stem-leaves opposite or 3–4 in a whorl, rounded, entire, net-veined, fleshy, stalked except the uppermost. Flowers white, *c.* 5 mm across, on slender stalks in lax rounded terminal cluster 2–3.5 cm across. Sepals short, blunt. Petals narrow, acute.

C. Nepal to S.W. China. 2000–2400 m. Aug.–Sep. **Pl. 39.**

180 **Sedum multicaule** See F.H. p. 143, p. 468. **Pl. 38.**

181 **Sedum oreades** See F.H. p. 144, p. 468. **Pl. 39.**

COMBRETACEAE

182 **Combretum roxburghii** Spreng. (*C. decandrum* Roxb.)
Large climber festooning forest trees, conspicuous when in flower because of its large white leaf-like bracts. Leaves opposite, elliptic-oblong long-pointed, nerves prominent beneath. Flowers greenish-white, *c.* 4 mm across, in dense hairy spikes *c.* 2.5 cm long, forming large terminal clusters in the axils of leaf-like creamy bracts 2.5–5 cm long. Fruit oblong-elliptic, with 5 papery wings.

Himachal Pradesh to S.W. China. To 600 m. (S.E. Asia) Dec.–Mar. **Pl. 40.**

MYRTACEAE Myrtle Family

183 **Callistemon citrinus** BOTTLE-BRUSH TREE See F.H. p. 145. **Pl. 41.**

LYTHRACEAE Loosestrife Family

184 **Rotala rotundifolia** See F.H. p. 147, p. 468. **Pl. 40.**

SONNERATIACEAE

185 **Duabanga grandiflora** (Roxb. ex DC.) Walp. (*D. sonneratioides* Buch.–Ham.)

Evergreen tree to 30 m with pendulous branches. Leaves opposite, large, to 25 cm, oblong acute, entire, cordate or rounded at base. Flowers white, 5–7.5 cm across, in large pendulous terminal clusters. Calyx cup-shaped, deeply 6–8–lobed; lobes thick, rigid, ultimately spreading. Petals obovate, clawed, margins crisped and undulate. Stamens longer than petals. Capsule globose, on persistent spreading calyx.

Uttar Pradesh to S.W. China. 250–1100 m. (Burma, S.E. Asia) Feb.– Mar. **Pl. 40.**

CUCURBITACEAE   Gourd or Cucumber Family

186 **Trichosanthes lepiniana** (Naudin) Cogn.
Robust climber with grooved branched hairless stem, and with usually 3–fid tendrils. Leaves palmately 3–5–lobed, cordate at base, bright green and with translucent glands above, paler and prominently net-veined beneath; lobes of leaf broadly ovate or triangular, acute at apex, margins remotely toothed; leaf-stalk grooved, with translucent glands. Flowers white, 4–6 cm across. Male flowers usually in paired clusters, female flowers solitary. Calyx-tube longitudinally grooved, narrowing from apex to base, with toothed lobes. Petals 2–3 cm long, obovate, deeply cut and fringed. Bracts many-nerved, downy, fringed with long awl-shaped teeth. Fruit red, smooth, ovoid, 8 cm long by 6 cm broad.

Uttar Pradesh to Sikkim. 300–2100 m. May–Jul. **Pl. 40.**
Further work has recently been done on this difficult genus. Almost all the material from Nepal previously named as *T. tricuspidata* has now been renamed as *T. lepiniana*. It seems probable that the plant shown on Pl. 48 of F.H. was *T. lepiniana*, for of the many gatherings from Nepal only one, from the extreme east of the country, remains named as *T. tricuspidata*. See C. Jeffrey. Further Notes on Cucurbitaceae, *Kew Bull.* 1982, Vol. 36(4), p. 737.

CACTACEAE   Cactus Family

187 **Cereus peruvianus** (L.) Mill.
Erect succulent leafless prickly cactus to 4 m, with dull bluish-green parallel ribbed branches. Ribs compressed, wavy, armed with prickles in tufts of 8–20. Flowers white, solitary. Sepals and petals forming a funnel-shaped perianth to 16 cm long and 8 cm across.

Native of C. America. Sometimes used as a hedge-plant in Himalaya. To 1000 m. May–Jun. **Pl. 41.**

UMBELLIFERAE   Umbellifer Family

188 **Eryngium billardieri** Delaroche.
Spiny bluish perennial of wasteland in Kashmir valley, with erect branched stems to 70 cm. Leaves pinnately divided, margins of lobes spiny; basal leaves stalked, stem-leaves stalkless. Flower-heads to 1.5 cm across, subtended by an involucre of bracts alternating with spines; margins of bracts entire or spiny. Bractlets linear, longer than flowers. Calyx-tube with white scales; calyx-lobes lanceolate-ovate, with long needle-like tips. Petals notched. Fruit ellipsoid, scaly.
   Afghanistan to Kashmir. 1500 m. (W. Asia) Jun.–Sep. **Pl. 41.**

Eryngium biebersteinianum Very similar to 188, but distinguished by its undivided basal leaves. See F.H. p. 155, p. 469.

189 **Heracleum nepalense** See F.H. p. 158. **Pl. 41.**

190 **Heracleum wallichii** See F.H. p. 158. **Pl. 41.**

ARALIACEAE   Ivy Family

191 **Acanthopanax cissifolius** See F.H. p. 161, p. 470. **Pl. 42.**

192 **Aralia cachemirica** See F.H. p. 161, p. 470. **Pl. 43.**

193 **Brassaiopsis mitis**, C.B. Clarke
   Evergreen tree to 6 m in understory of dense forests. Young parts and inflorescence densely covered with bristly hairs. Leaves crowded towards ends of prickly branches, deeply palmately lobed; lobes oblong-lanceolate long-pointed, narrowed towards base, toothed, softly hairy beneath; leaf-stalks bristly above. Stipules large, lanceolate-awl-shaped. Flowers *c.* 7 mm across, in umbels 3–4 cm across arranged in a large branched inflorescence. Bracts lanceolate, soon falling. Calyx stellately hairy when young. Petals white, hairless. Disk large, hemispheric. Fruit globose.
   E. Nepal to Sikkim. 1200–2400 m. Jun.–Jul. **Pl. 42.**

194 **Hedera nepalensis** See F.H. p. 160.
Ripe fruit is usually bright yellow, but sometimes orange-red. **Pl. 42.**

195 **Helwingia himalaica** See F.H. p. 161, p. 471. **Pl. 44.**

196 **Merrilliopanax alpinus** (C.B. Clarke) Shang (*Brassaiopsis a.* C.B. Clarke, *Tetrapanax tibetanus* Hoo)
   Unarmed shrub or small tree, with branchlets covered with dense brownish stellate hairs. Leaves oval, oval-oblong or almost circular, cordate or rounded at base, unlobed or with 3 oval-triangular lobes, long-pointed at apex, irregularly toothed, both surfaces at first densely covered with stellate grey-brown hairs, becoming hairless except on nerves beneath;

main nerves 5–7, digitate from base; leaf-stalk dilated at base. Flowers yellow-green, in umbels 1 cm across, forming a large branched densely stellate-hairy inflorescence to 30 cm long. Calyx with 5 small triangular teeth. Petals 5, oblong-triangular, 2 mm long, reflexed. Fruit black, globose, 3–4 mm.

E. Nepal to S.W. China. 1500–3100 m. May–Jun. Pl. 43.

197 **Pentapanax leschenaultii** See F.H. p. 162, Pl. 49. Pl. 43.

198 **Schefflera impressa** (C.B. Clarke) Harms (*Heptapleurum i.* C.B. Clarke)

Unarmed tree to 15 m. Leaves digitate, stalked; leaflets lanceolate long-pointed, entire or remotely toothed, nerves deeply impressed above, stellately woolly beneath. Flowers maroon, in stellately woolly umbels forming a large inflorescence 30–38 cm across. Fruit globose, 5–6-angled.

Uttar Pradesh to S.W. China. 2000–3400 m. Jul.–Aug. Pl. 42.

199 **Schefflera venulosa** (Wight & Arn.) Harms (*Heptapleurum v.* (Wight & Arn.) Seem.)

Unarmed evergreen climbing shrub on rocks and trees. Leaves digitate, with leaf-stalk dilated and stem-clasping at base; leaflets elliptic-oblong long-pointed, hairless, leathery, shining above, stalked. Flowers cream-coloured, in globular umbels *c.* 1.5 cm across, arranged in large branched purple-stemmed clusters. Calyx truncate (cut-off at top). Petals 5, obovate-oblong acute. Fruit orange-yellow, globose, 5–ridged.

Himachal Pradesh to S.W. China. 300–1800 m. Jan.–May. Pl. 43.

200 **Trevesia palmata** See F.H. p. 162, p. 470. Pl. 42.

CORNACEAE Dogwood Family

201 **Swida oblonga** See F.H. p. 163, p. 471. Pl. 44.

ALANGIACEAE

202 **Alangium alpinum** (C.B. Clarke) W.W. Smith & Cave (*Marlea begoniaefolia* Roxb. var. *a.* C.B. Clarke)

Tree to 12 m or more. Leaves alternate, orbicular to broad-oblong, long-pointed, entire, base rounded or cordate, hairy beneath. Flowers white, in few-flowered branched axillary clusters. Calyx-tube shortly toothed. Petals 6–8, linear, 1.5–2 cm long, recurved. Fruit ovoid, obscurely ribbed, hairless, crowned with calyx.

C. Nepal to Bhutan. 1900–2700 m. (Burma) May–Jun. Pl. 44.

203 **Alangium salviifolium** (L.f.) Wangerin (*A. lamarckii* auct. non Thwaites p.p.)

Shrub or small tree to 5 m, flowering when almost leafless. Leaves oblong-elliptic, acute or blunt, base unequal, with gland-pits or tufts of hairs in axils of nerves beneath, and with woolly-hairy leaf-stalks. Flowers creamy-white, solitary or in close axillary clusters, all parts woolly-hairy. Calyx cup-shaped, shortly toothed. Petals 1.25–2.8 cm long, oblong-acute, recurved. Fruit a succulent black ellipsoid berry crowned with calyx. Bark and roots have medicinal uses.

Uttar Pradesh to E. Nepal. To 350 m. (India, China, S.E. Asia, Africa) Feb.–Apr. **Pl. 44.**

CAPRIFOLIACEAE   Honeysuckle Family

204 **Lonicera angustifolia**   See F.H. p. 167. **Pl. 45.**

205 **Lonicera hypoleuca**   See F.H. p. 168. **Pl. 45.**

206 **Lonicera lanceolata**   See F.H. p. 167. **Pl. 45.**

207 **Lonicera ligustrina** Wallich
Shrub to 3 m. Leaves glossy, lanceolate long-pointed, hairy on midrib beneath, with short hairy leaf-stalk. Calyx tubular. Corolla yellow or white, hairy; corolla-tube 6 mm long, narrowly funnel-shaped, humped at base, with blunt lobes 4–5 mm. Bracteoles fused at base, enlarged in fruit to form boat-shaped cover half-enclosing the small spherical bright red juicy berries.
C. Nepal to S.W. China. 1800–2100 m. (India, Burma) Apr.–May. **Pl. 46.**

208 **Lonicera tomentella** Hook.f. & Thoms.
Shrub to 5 m, with downy branchlets. Leaves oblong, blunt at both ends, hairy beneath and with hairy leaf-stalks. Bracts leafy, narrowly oblong. Flowers pink or white, paired, on short stalks. Bracteoles fused at base. Calyx hairless. Corolla tubular, 6 mm long, with lobes 2–3 mm, hairy. Distinguished from other species by corolla being neither 2–lipped nor humped at base. Berries black, often coalescing in pairs into one spherical fruit.
C. Nepal to Bhutan. 2400–3600 m. (Tibet) Jun.–Aug. **Pl. 45.**

SAMBUCACEAE   Elder Family

209 **Sambucus adnata**   See F.H. p. 169, p. 472. **Pl. 46.**

210 **Sambucus wightiana**   See F.H. p. 169. **Pl. 46.**

211 **Viburnum colebrookianum** Wallich ex DC.
Spreading shrub 2–4 m, with young parts stellately hairy. Leaves somewhat drooping, elliptic-oblong, saw-toothed, with arching nerves. Flowers

27

white, in dense erect compound axillary umbels. Calyx tubular, with minute lobes, hairless. Corolla-lobes rounded, *c.* 1.5 mm. Fruit red, orbicular.

E. Nepal to Bhutan. To 900 m. (S.E. Asia) Feb.–Mar. **Pl. 46.**

212 **Viburnum cylindricum** See F.H. p. 170, p. 473. **Pl. 44.**

213 **Viburnum mullaha** See F.H. p. 171, p. 473. **Pl. 45.**

RUBIACEAE Madder Family

214 **Asperula oppositifolia** Regel & Schmalh. subsp. **pseudo-cynanchica** Ehrend.
Tuft-forming perennial of rocks and cliffs, with stout rootstock and many erect or ascending slender stems to 25 cm. Leaves small linear, often recurved; lower leaves in whorls of 4, upper leaves paired. Flowers white or pink, in terminal clusters subtended by 2 small linear-oblong acute bracts. Corolla-tube funnel-shaped, *c.* 2.5 mm long, with short triangular lobes. Plants similar to the one illustrated are quite common in Miyar Nullah in Lahul, and also in adjacent parts of the Chenab valley. Specimens from here have only provisionally been named as this species, and further study on the genus is needed.
Pakistan to Himachal Pradesh. 1500–3000 m. May–Jun. **Pl. 47.**

215 **Galium verum** LADY'S BEDSTRAW See F.H. p. 175. **Pl. 47.**

216 **Mussaenda roxburghii** Hook.f.
Shrub to 3 m, with stem usually hairy above. Leaves oblong-lanceolate long-pointed, bristly-hairy on midrib beneath, short-stalked. Flowers in dense terminal clusters *c.* 5 cm across. Calyx-lobes 5; one lobe often enlarged into a prominent white oblanceolate stalked 'leaf'. Corolla-tube narrow, 3–3.5 cm long, silky; lobes bright orange inside, white outside, with short pointed tips. Berries oblong-ellipsoid, hairless, crowned with the calyx-lobes.
C. Nepal to Bhutan. To 1600 m. (Assam, Burma) May–Aug. **Pl. 47.**

217 **Wendlandia coriacea** (Wallich) DC.
Hairless shrub to 5 m. Leaves opposite, leathery, shining above, narrowly lanceolate long-pointed, stalked. Flowers white, *c.* 8 mm long, stalkless, in dense branched inflorescence. Calyx hairless, with short rounded teeth. Corolla-tube slender, funnel-shaped, with broad-oblong lobes shorter than tube. Stamens shortly exserted. Capsule globose.
W. Nepal to Sikkim. To 1900 m. Mar.–Apr. **Pl. 47.**

COMPOSITAE Daisy Family

218 **Ageratum conyzoides** See F.H. p. 181, p. 475. **Pl. 48.**

219 **Anaphalis busua** See F.H. p. 187. **Pl. 48.**

220 **Anthemis cotula** STINKING MAYWEED See F.H. p. 190. **Pl. 49.**

221 **Arctium lappa** L. GREAT BURDOCK
Erect branched biennial to 1 m or more. Leaves ovate-cordate, margins undulate, usually cottony beneath, stalked. Flower-heads globose, 1.8–3.7 cm across, in terminal clusters. Involucral bracts with rigid awl-like spreading barbed tips. Corolla and stamens purple-pink, styles white. Achenes oblong, ribbed, angled.
Pakistan to C. Nepal. 2000–4000 m. (Temperate Eurasia) Jun.–Aug. **Pl. 49.**

222 **Artemisia absinthium** WORMWOOD See F.H. p. 194. **Pl. 48.**

223 **Artemisia dracunculus** TARRAGON See F.H. p. 194. **Pl. 48.**

224 **Artemisia gmelinii** See F.H. p. 195, p. 476. **Pl. 48.**

225 **Aster indamellus** See F.H. p. 182. **Pl. 50.**

226 **Aster molliusculus** See F.H. p. 184, p. 475. **Pl. 49.**

227 **Aster sikkimensis** Hook.
Erect leafy perennial to 1 m, with numerous flexuous branches. Leaves lanceolate long-pointed, toothed, thin. Flowers purplish-pink, in heads 15 mm across borne in flat-topped clusters. Involucral bracts linear long-pointed. Achenes hairy, 4–ribbed. Pappus white or reddish.
E. Nepal to Sikkim. 2400–3000 m. Oct.–Nov. **Pl. 49.**

228 **Aster thomsonii** See F.H. p. 184. **Pl. 50.**

229 **Aster trinervius** See F.H. p. 182. **Pl. 50.**

230 **Carduus edelbergii** See F.H. p. 208, p. 479. **Pl. 50.**

231 **Centaurea iberica** See F.H. p. 211, p. 480. **Pl. 49.**

232 **Chrysanthemum pyrethroides** See F.H. p. 193, p. 475. **Pl. 51.**

233 **Cichorium intybus** CHICORY See F.H. p. 213. **Pl. 50.**

234 **Crassocephalum crepidioides** (Benth.) S. Moore
Annual weed of cultivated areas, with hairy erect simple or branched stem to 80 cm. Leaves ovate, acutely irregularly toothed and sometimes lobed, stalked. Flowers orange-red, in drooping narrow-oblong heads 1–1.5 cm long, forming a dense or lax cluster. Involucral bracts linear, with awl-shaped tips, more or less hairless. Pappus white.
Native of America, naturalised throughout the Tropics. C. Nepal to Sikkim. 400–1900 m. Flowers most of year. **Pl. 50.**

235 **Cremanthodium purpureifolium** See F.H. p. 197. **Pl. 52.**

236 **Crepis flexuosa** See F.H. p. 217, p. 481. **Pl. 51.**

237 **Crepis sancta** See F.H. p. 218. **Pl. 51.**

238 **Cyathocline purpurea** (Buch.–Ham. ex D. Don) O. Kuntze (*C. lyrata* Cass.)

Erect aromatic annual weed of fallow paddy-fields, with leafy sparsely hairy branched stem to 60 cm. Leaves pinnately lobed; lobes toothed. Flowers rose-purple, in hemispheric heads *c.* 4 mm across borne in small rounded hairy clusters. Ray-florets nil. Involucral bracts lanceolate acute, with papery margins. Achenes minute, oblong, smooth. Pappus nil.

Pakistan to Bhutan. 600–1300 m. (India, China, S.E. Asia) Oct.–Apr. **Pl. 51.**

239 **Dubyaea hispida** See F.H. p. 216, p. 481. **Pl. 52.**

240 **Eupatorium odoratum** L.

Erect softly hairy more or less aromatic perennial to 2 m, with many spreading branches, gregarious on wasteland and often forming dense thickets. Leaves rhomboid-ovate or ovate-lanceolate, long-pointed, base wedge-shaped, with 3 main nerves, margins toothed, densely hairy beneath, stalked. Flowers pale pinkish-mauve, in cylindric heads 1 cm long arranged in branched terminal clusters. Involucral bracts oblong blunt, white lined with green, margins papery. Achenes rough, 5–sided. Pappus white.

Native of America. Widely naturalised in C. & E. Nepal. To 1500 m. Dec.–Apr. **Pl. 53.**

241 **Gnaphalium affine** See F.H. p. 185, p. 475. **Pl. 53.**

242 **Gynura cusimbua** See F.H. p. 198, p. 476. **Pl. 52.**

243 **Gynura nepalensis** See F.H. p. 199. **Pl. 52.**

244 **Jurinea ceratocarpa** Benth. var. **depressa** C.B. Clarke

More or less stemless perennial of drier areas, with thick rootstock. Leaves linear-oblong or lanceolate, entire shallow-lobed or deeply pinnate-lobed, grey-green above, densely white-woolly beneath. Flower-heads pinkish-purple, *c.* 1 cm across, in a central cluster surrounded by an involucre of leaves. Involucral bracts narrow-lanceolate long-pointed; inner bracts narrower and purple-tipped. This var. is very different from the normal form of the species (see F.H. p. 208, p. 479).

Pakistan to Himachal Pradesh. 3600–4100 m. Aug. **Pl. 53.**

245 **Jurinea sp.**

The plant illustrated has not been named. Dried material collected at the time the photo was taken indicate that its nearest relative is *J. berardioides* of Baluchistan and W. Asia, but it differs from that species in several respects. It grew in Kashmir near the track which leads from Baltal to the pilgrimage cave at Amarnath. 3300 m. July. **Pl. 53.**

246 **Leontopodium monocephalum** Edgew.

Much-branched mat-forming woolly perennial scree-plant, with many stolons bearing tight rosettes of leaves. Lower leaves spathulate, densely

white-woolly; involucral leaves densely yellow-woolly, surrounding flower-heads 5–10 mm across. Involucral bracts lanceolate acute, margins hair-fringed and broadly papery. Achenes 4–ribbed. Pappus white.
Pakistan to Sikkim. 4600–5600 m. Jul.–Oct. **Pl. 54.**

247 **Leontopodium stracheyi** (Hook.f.) C.B. Clarke ex Hemsley
Tufted glandular-hairy perennial with leafy unbranched stems 20–35 cm. Stem-leaves narrow-lanceolate or sometimes linear, apex acute, base stem-clasping, margins wavy-toothed, bright green above, white-woolly beneath. Flower-heads in dense domed woolly terminal clusters 1–2.5 cm across, each cluster surrounded by an involucre of uppermost lanceolate densely white-woolly leaves 1–2 cm long. Easily distinguishable from the other *c.* 9 species of this very difficult genus occurring in our area both by its general appearance and by its glandular-hairy stems.
Uttar Pradesh to S.W. China. 2200–4500 m. Jul.–Oct. **Pl. 54.**

248 **Ligularia fischeri** See F.H. p. 202, p. 477.
This is a very variable species, which in its original description was published as a *Senecio*. It is rather confusing that the closely related Kashmir plant should still strictly be called *Senecio jacquemontianus*, despite the fact that botanists generally regard it to be a *Ligularia*. Presumably at some future date this latter species will be republished as a *Ligularia*, and perhaps even reduced to a variety of *L. fischeri*. **Pl. 56.**

249 **Nannoglottis hookeri** See F.H. p. 199, p. 477. **Pl. 54.**

250 **Petasites tricholobus** See F.H. p. 196. **Pl. 56.**

251 **Saussurea atkinsonii** C.B. Clarke
Stemless perennial of drier areas with undivided rootstock. Leaves in procumbent rosette, stalked or stalkless, broadly obovate-elliptic, coarse-ly toothed, prominently veined. Flowers dull purple, in solitary central flower-head to 2.5 cm long. Inner involucral bracts linear-oblong, papery; outer bracts variable, erect or recurved. Achenes hairless. Pappus brown, feathery.
Pakistan to Kashmir. 3000–4500 m. Aug.–Sep. **Pl. 55.**

252 **Saussurea bracteata** See F.H. p. 207. **Pl. 55.**

253 **Saussurea costus** See F.H. p. 207, p. 478. **Pl. 55.**

254 **Senecio cappa** See F.H. p. 200. **Pl. 57.**

255 **Senecio diversifolius** See F.H. p. 201. **Pl. 57.**

256 **Senecio triligulatus** See F.H. p. 200. **Pl. 57.**

257 **Senecio wallichii** See F.H. p. 200. **Pl. 57.**

258 **Soroseris pumila** See F.H. p. 216. **Pl. 55.**

259 **Sphaeranthus indicus** L.
Annual weed of fallow paddy-fields, with erect or spreading softly hairy stems 15–45 cm and winged branches. Leaves obovate-oblong, narrowed

to decurrent base, toothed. Flower-heads tiny, crowded into purple-pink globose solitary terminal heads *c.* 1 cm across. Involucral bracts linear, hairy, nearly as long as flowers. The plant has medicinal uses.

Himachal Pradesh to Bhutan. To 1000 m. (India, S.E. Asia) Dec.–Apr. Pl. 52.

260 **Tagetes patula** L. FRENCH MARIGOLD
Hairless annual with branching stem to 50 cm. Leaves stalkless, pinnately cut into usually 5–7 pairs of oblong-lanceolate sharply toothed lobes bearing scattered glands. Flower-heads 3–4 cm across, solitary at ends of branches on stalks which are swollen below the head. Involucre blue-green, cylindric, with 5 short triangular teeth. Ray-florets rounded, notched at apex, *c.* 1 cm, yellow or orange, often with reddish tinge. Disk-florets with style-arms exserted.

Native of America. Grown in gardens throughout Himalaya, and naturalized around villages. Much used to decorate shrines. 900–2000 m. Apr.–Oct. Pl. 56.

261 **Tanacetum nubigenum** See F.H. p. 192. Pl. 51.

262 **Vernonia volkameriifolia** DC.
Small tree with downy branchlets. Leaves obovate-oblong or lanceolate, acute or blunt, entire or irregularly toothed, with leaf-stalk half-encircling stem at base. Flowers pale mauve, in heads *c.* 8 mm long, very numerous in a large branched terminal inflorescence. Involucral bracts oblong, with blunt or rounded apex, hairy. Achenes 10–ribbed. Pappus whitish.

C. Nepal to S.W. China. 400–2400 m. (S.E. Asia) Feb.–Apr. Pl. 56.

263 **Waldheimia nivea** (Hook.f. & Thoms. ex C.B. Clarke) Regel
Dwarf perennial densely clothed with white appressed wool. Leaves densely crowded, wedge-shaped, with 3–5 short blunt lobes. Flower-heads stalkless, 12–18 mm across, with 8–10 broadly elliptic pink or white ray-florets *c.* 6 mm long. Involucral bracts woolly, with narrow brown papery margins. Achenes glandular. Pappus yellowish, with slightly dilated tips.

Pakistan to C. Nepal. 3900–5400 m. (C. Asia) Jun.–Aug. Pl. 54.

264 **Youngia tenuifolia** Babcock & Stebb. subsp. **diversifolia**
(Ledeb. ex Spreng.) Babcock & Stebb. (*Crepis t.* Willd.)
Hairless perennial of Tibetan borderlands, with stout rootstock and many stiff much-branched stems to 50 cm, clothed at base with remains of old leaves. Basal leaves pinnately lobed, with winged rhachis and long leaf-stalk; lobes unequal, entire or sparingly toothed. Stem-leaves usually linear, entire. Flower-heads yellow, 1.25 cm long, in branched flat-topped clusters. Outer involucral bracts very short; inner much longer, hairy, linear acute, tips often clawed. Achenes narrowly spindle-shaped, ribbed, beaked. Pappus silvery.

Pakistan to C. Nepal. 3000–4600 m. (Tibet, C. Asia) Jun.–Aug. Pl. 54.

CAMPANULACEAE Bellflower Family

265 **Codonopsis purpurea** See F.H. p. 222. Pl. 59.

266 **Codonopsis thalictrifolia** See F.H. p. 222. Pl. 58.

267 **Cyananthus microphyllus** See F.H. p. 223. Pl. 58.

268 **Lobelia seguinii var. doniana** See F.H. p. 224, p. 482. Pl. 58.

ERICACEAE Heath Family

269 **Agapetes serpens** For the usual red-flowered epiphytic form see F.H. p. 235, Pl. 77. A form with white or pale yellow flowers is also common in parts of E. Nepal, growing usually as a small erect terrestrial undershrub. Pl. 59.

270 **Gaultheria fragrantissima** See F.H. p. 235, Pl. 77. Pl. 59.

271 **Gaultheria griffithiana** See F.H. p. 235. Pl. 59.

272 **Gaultheria hookeri** See F.H. p. 235. Pl. 60.

273 **Gaultheria trichophylla** See F.H. p. 234, Pl. 77. Pl. 58.

274 **Rhododendron arboreum** For the usual medium-altitude form with blood-red flowers see F.H. p. 228, Pl. 74. At higher altitudes the flowers are often pale pink or white. Pl. 61.

275 **Rhododendron campanulatum** For the usual mauve-flowered form see F.H. p. 226, Pl. 74. The white-flowered form, which is not common, is much more handsome. Pl. 61.

276 **Rhododendron campylocarpum** See F.H. p. 226. Pl. 60.

277 **Rhododendron fulgens** See F.H. p. 227. Pl. 61.

278 **Rhododendron grande** See F.H. p. 227. Pl. 61.

279 **Rhododendron lanatum** Hook.f.
Shrub 1–4 m, often gregarious and forming alpine thickets. Branchlets clothed with soft white or tawny felted wool. Leaves obovate-elliptic, acute or blunt, hairless above except in groove of midrib, densely woolly beneath and on leaf-stalk. Flowers in small trusses of 6–10. Calyx *c.* 2 mm long, with 5 broad hairless lobes. Corolla broadly campanulate, *c.* 4.5 cm long, pale sulphur-yellow, red-spotted within; lobes 5, broad rounded, *c.* 1.5 cm by 2 cm. Capsule oblong cylindric, slightly curved, tawny-woolly. It seems that this species has not been recorded from Nepal, but it may well occur there because it grows abundantly nearby on the east side of the Singalelah Ridge in Sikkim.
    Sikkim to Arunachal Pradesh. 3000–4000 m. May. Pl. 62.

280 **Rhododendron pumilum** See F.H. p. 230, p. 483. Pl. 62.

281 **Vaccinium dunalianum** Wight
Erect evergreen forest undershrub to 3 m, sometimes epiphytic, with angular uniformly leafy branches. Leaves leathery, oblong-lanceolate long-pointed, entire, with recurved margins. Flowers pinkish-green in axillary clusters. Bracts densely overlapping, round-ovate, with papery margins. Calyx-teeth small triangular. Corolla ovoid-conic, 5 mm, hairless outside, with small recurved lobes. Berry black, globose.
C. Nepal to S.W. China. 1600–3000 m. (Burma, S.E. Asia) May. Pl. 59.

282 **Vaccinium retusum** See F.H. p. 236. Pl. 60.

283 **Vaccinium sikkimense** C.B. Clarke
Small rigid shrub to 1 m, with softly hairy branchlets. Leaves obovate-oblong acute, finely toothed, leathery, hairless, stalkless. Flowers pink, 5 mm, in short dense terminal clusters. Bracts elliptic. Calyx-tube globose, with 5 short nearly hairless teeth. Corolla ovoid, deeply lobed. Berry globose, when immature green with crimson ring at apex. Rare in our area.
E. Nepal to S.W. China. 3000–4000 m. May. Pl. 60.

PYROLACEAE Wintergreen Family

284 **Pyrola karakoramica** See F.H. p. 237, p. 484.
At one time this attractive little plant was not uncommon in Kashmir on the route from Sonamarg to the Vishensar and Gangabal lakes. It seems to have disappeared there in recent years. Apart from Kashmir, the genus is recorded in our area only from the extreme east. The E. Himalayan *Pyrola sikkimensis* has been found at Kambachen near Kanchenjunga in E. Nepal. Pl. 58.

DIAPENSIACEAE

285 **Diapensia himalaica** See F.H. p. 237, p. 484. Pl. 58.

PRIMULACEAE Primula Family

286 **Androsace lanuginosa** See F.H. p. 240. Pl. 62.

287 **Androsace muscoidea** See F.H. p. 241. Pl. 62.

288 **Androsace muscoidea** Duby forma **longiscapa** (Knuth) Hand.–Mazz.
Perennial of drier areas, with brown stolons and fibrous roots, forming loose mats. Leaves oblanceolate, incurving, in tight globular silky-haired

rosettes. Flowers *c*. 10 mm across, mauve-pink, with yellow eye later turning red, 2–8 in silky-haired umbels on erect hairy stems to 6 cm. Bracts lanceolate, incurving. Corolla-tube globular, with rounded spreading lobes. This is a very distinct form, and should perhaps more correctly be called *Androsace robusta*. See Smith & Lowe (1977), *Androsaces*, Alpine Garden Soc.

Pakistan to C. Nepal. 3100–5600 m. (W. Tibet) Jun.–Jul. Pl. 62.

289 **Androsace nortonii** Ludlow
Tiny tufted stoloniferous perennial forming loose mats in drier areas. Leaves in tight rosettes, densely covered with silky white hairs; external leaves narrowly elliptic; intermediate leaves strap-shaped or spathulate, with blunt apices; inner leaves stalked, with elliptic blunt blades. Flowering stem 1–4 cm, slender, hairy, bearing umbel of 2–6 flowers. Bracts linear, hairy. Calyx densely hairy, 5–lobed to middle; lobes narrowly ovate, 3–nerved. Corolla *c*. 5 m across, rose-pink, with red or green eye; corolla-lobes broadly obovate. Resembling *Androsace villosa*, but distinguished by its 3 kinds of leaf.

C. to E. Nepal. 3500–4700 m. (Tibet) Jun.–Jul. Pl. 62.

290 **Primula barnardoana** W.W. Smith & Ward
Rotundifolia section. Rhizome short, clothed at base with farinose overlapping ovate bud-scales. Leaves ovate, rounded or blunt at apex, base cordate or rounded, toothed, often creamy-farinose beneath; leaf-stalk winged, sheathing at base. Flowering stem 10–40 cm, bearing umbel of yellow flowers, or occasionally 2 superposed umbels. Bracts lanceolate long-pointed. Calyx tubular, 7–10 mm, conspicuously 5–ribbed, with lanceolate acute lobes. Corolla *c*. 1.7 cm across, yellow-orange or yellow-green, with darker yellow eye; tube cylindric, twice as long as calyx; lobes obovate, entire or slightly notched. The plant illustrated was grown at the Royal Botanic Garden Edinburgh from seed collected in E. Nepal.

E. Nepal to S.E. Tibet. 3600–4500 m. May–Jul. Pl. 63.

291 **Primula boothii** Craib
Petiolares Section. Efarinose perennial with short stout rhizome devoid of basal bud-scales at flowering-time. Leaves in a rosette, coarsely and irregularly toothed, midrib and leaf-stalk tinged red; outer leaves elliptic, strongly stalked; inner leaves spathulate-elliptic, almost stalkless. Flowers lilac-mauve, 1.5–2.5 cm across, with yellow eye, on flowering stem to 5 cm or nil. Calyx downy, campanulate, with narrow-triangular pointed lobes. Corolla-tube cylindric, at least twice as long as calyx; lobes obovate, often 3–toothed.

E. Nepal. to S.E. Tibet. 2400–3400 m. Apr.–May. Pl. 63.

292 **Primula buryana** See F.H. p. 251. Pl. 66.

293 **Primula buryana** Balf.f. var. **purpurea** Fletcher
Whereas the white-flowered form of *Primula buryana* grows in drier areas

adjacent to the Tibetan border, the blue-flowered form grows on rock-ledges and peat-banks in wet country fully exposed to the monsoon rains. The flowers of the latter are often solitary, or 2–3 in a much laxer head than that of the white-flowered form.

C. Nepal endemic. 2900–4600 m. Jun.–Jul. Pl. 63.

294 **Primula capitata** Hook.f. subsp. **crispata** (Balf.f. & W.W. Smith) W.W. Smith & Forrest.
Capitatae Section. Perennial, devoid of basal bud-scales at flowering time. Leaves efarinose, oblong-spathulate, acute or rounded at apex, tapering to winged leaf-stalk, toothed; midrib, lateral nerves and veining prominent. Flowering stem to 30 cm, farinose above, bearing a disk-shaped head of crowded deflexed stalkless purple-blue flowers *c.* 8 mm long by 5 mm across which open in succession, the upper unopened ones depressed and overlapping. Calyx white-farinose, with broadly ovate-elliptic lobes. Corolla-tube cylindric, with obcordate notched spreading lobes. Distinguished from other subspecies by its efarinose leaves. Distinguished from the rather similar *Primula glomerata* (See F.H. p. 243, Pl. 82) by its drooping flowers (*P. glomerata* flowers more or less erect).

E. Nepal to S.E. Tibet. 2800–4300 m. (Burma) Jun.–Sep. Pl. 67.

295 **Primula concinna** Watt
Farinosae Section. Dwarf densely tufted plant with very short rhizome. Leaves oblanceolate, spathulate or obovate, narrowed to winged leaf-stalk, entire or obscurely rounded-toothed, usually hairless above, thickly yellow-farinose below. Flowering stem 1–2.5 cm, much shorter than the individual flower-stalks. Bracts linear-oblong. Calyx campanulate, more or less farinose, lobed to the middle. Corolla pale pink, mauve or white, 4–8 mm across, with yellow eye; tube short, equalling calyx; lobes narrowly obovate, notched.

W. Nepal to S.E. Tibet. 4000–5000 m. Jun.–Aug. Pl. 63.

296 **Primula deuteronana** See F.H. p. 249. Pl. 65.

297 **Primula dickieana** Watt
Amethystina Section. Rootstock clothed above with numerous small scale-leaves. Leaves in a tufted basal rosette, elliptic-obovate to oblanceolate, more or less acute at apex, wedge-shaped below, margins entire or remotely toothed, efarinose, with short winged leaf-stalk. Flowering stem 8–20 cm, bearing umbel of 1–6 flowers 2–3 cm across, pale yellow with deeper yellow eye. Bracts linear-thread-like. Corolla-tube twice as long as calyx, hairy at mouth and within; corolla-lobes spreading, obcordate-elliptic-oblong, deeply bilobed at apex.

E. Nepal to S.W. China. 4000–5000 m. Jun.–Aug. Pl. 66.

298 **Primula drummondiana** Craib
Petiolares Section. Rhizome short, devoid of basal bud-scales at flowering time. Leaves spathulate, oblanceolate or obovate, rounded at apex,

narrowed to broadly winged leaf-stalk, irregularly toothed, efarinose, glandular-hairy above. Flowers *c.* 1.5 cm across, mauve with yellow eye, on very slender stalks. Bracts lanceolate. Calyx 5–7 mm, tubular-campanulate, with lanceolate acute lobes and little or no farina. Corolla-tube cylindric, longer than calyx, with narrowly obcordate deeply notched lobes. The plant illustrated was grown in England by Mr J. Templer from seed collected from Annapurna Himal.

Uttar Pradesh to C. Nepal. 3300–4000 m. Jun.–Sep. **Pl. 63.**

299 **Primula floribunda** See F.H. p. 246. **Pl. 67.**

300 **Primula glabra** Klatt
Farinosae Section. Leaves in compact rosette, oblong-spathulate, rounded or blunt at apex, margins with teeth often recurved, hairless, efarinose, minutely gland-dotted beneath, tapering to winged leaf-stalk. Flowers 4–7 mm across, pinkish-purple to bluish-violet, with yellow or orange eye, in clusters on flowering stem 2–11 cm. Calyx campanulate, sometimes unequally swollen at base, minutely glandular-downy, with ovate-oblong rounded lobes. Corolla-tube equalling or slightly exceeding calyx, with obovate deeply notched lobes.

E. Nepal to S.E. Tibet. 3800–4800 m. Jun.–Aug. **Pl. 64.**

301 **Primula glandulifera** Balf.f. & W.W. Smith
Minutissimae Section. Dwarf tufted plant covered at base with withered leaves; efarinose, but all parts except corolla covered with glandular down. Leaves elliptic, oblong-elliptic or wedge-shaped, rounded at apex, bluntly toothed, tapering to winged leaf-stalk. Flowers pale purplish-pink, with white eye, in a few-flowered cluster borne on stem 5 cm or less. Bracts strap-shaped, blunt at apex, broad and somewhat sheathing at base. Calyx cup-shaped, divided to well below middle into oblong-lanceolate blunt finely hairy lobes. Corolla-tube cylindric, to 1 cm; lobes spreading, 5 mm long, obovate, deeply notched.

Uttar Pradesh to C. Nepal. 3500–5600 m. May–Aug. **Pl. 64.**

302 **Primula gracilipes** See F.H. p. 250. **Pl. 63.**

303 **Primula megalocarpa** See F.H. p. 247. **Pl. 65.**

304 **Primula minutissima** See F.H. p. 246. **Pl. 64.**

305 **Primula muscoides** Hook.f. ex Watt
Minutissimae Section. Tiny tufted plant with base clothed with remains of withered leaves. Leaves efarinose, oblong-ovate, with 3–7 acute lobes at apex, stalkless. Flowers mauve with yellow eye, or pale violet with white eye, or not uncommonly pure white, solitary on very short stalks. Calyx yellow-farinose, cup-shaped, cleft to below middle into triangular lobes. Corolla-tube whitish, to 5 mm long, with white hairs in throat; corolla-lobes narrow-oblong, deeply notched.

C. Nepal. to S.E. Tibet. 4000–4900 m. Jun.–Aug. **Pl. 64.**

PRIMULACEAE

306 **Primula pulchra** Watt

Petiolares Section. Hairless efarinose plant of rock-ledges, with short thick rhizome. Bud-scales numerous at flowering-time. Leaves ovate-oblong, rounded or slightly cordate at base, margins wavy or rounded-toothed, blue-green below, somewhat fleshy. Flowers 1.5–2 cm across, purple or deep blue, with yellow eye, solitary or in an umbel borne on erect flowering stem. Bracts linear-lanceolate. Calyx tubular-campanulate, gland-dotted, divided almost to base into lanceolate lobes. Corolla-tube cylindric; lobes obcordate, irregularly notched.

E. Nepal to Sikkim. 3500–4500 m. May–Jul. Pl. 67.

307 **Primula ramzanae** Fletcher

Rotundifolia Section. Plant of rock-ledges and screes, with fairly stout rhizome. Leaves ovate-orbicular, cordate at base, deeply toothed, dark green and shining above, farinose beneath, long-stalked. Flowers fragrant, 2–2.5 cm across, in a loose umbel. Calyx divided almost to base into narrow-oblong lobes, densely farinose within. Corolla deep to pale purple, with yellow eye edged with white; tube 7 mm, twice the length of calyx; lobes obovate, deeply notched at apex.

W. Nepal endemic. Recorded only from a few localities near the Phoksumdo lake. 4300–4700 m. Jun. Pl. 65.

308 **Primula reidii** var. **williamsii**   See F.H. p. 252. Pl. 67.

309 **Primula sessilis**   See F.H. p. 249. Pl. 65.

310 **Primula sharmae**   See F.H. p. 244. Pl. 65.

311 **Primula sibirica**   See F. H. p. 245. Pl. 66.

312 **Primula soldanelloides** Watt

Soldanelloideae Section. Tiny hairless efarinose plant. Leaves 8–15 mm long, deeply pinnately divided, blunt at apex, tapering to leaf-stalk as long or longer than blade. Flowers white, nodding, solitary on slender stem 2.5–4 cm. Calyx dull blackish-green, campanulate, with triangular-ovate acute lobes. Corolla 10–15 mm long, broadly campanulate, with oblong-ovate lobes 3–5 mm long, notched or toothed at apex.

C. Nepal to S.E. Tibet. 3800–4600 m. Jul.–Aug. Pl. 64.

313 **Primula tenuiloba** (Watt) Pax

Minutissimae Section. Dwarf tufted efarinose plant covered at base with remains of withered leaves. Leaves spathulate, deeply toothed, blunt or rounded at apex, tapering at base to broad winged leaf-stalk much longer than blade. Flowers bluish-white with white eye, or sometimes pure white, 1.5–2 cm across, solitary on stem to 2 cm but often much shorter. Bracts solitary, linear, slightly pouched at base. Calyx to 5 mm long, campanulate, divided to middle into lobes strongly toothed at apex. Corolla-tube to 10 mm, white-hairy within and without; lobes obcordate, very deeply notched.

C. Nepal to S.E. Tibet. 3400–5200 m. Jun.–Aug. Pl. 64.

38

314 **Primula tibetica** See F.H. p. 245. **Pl. 65.**

315 **Primula uniflora** Klatt

Soldanelloideae Section. Leaves in a small rosette, ovate-spathulate, deeply toothed, white-hairy on both surfaces, efarinose; leaf-stalk about as long as blade. Flowering stem 7–15 cm, hairless, efarinose or almost so, bearing 1–2 nodding flowers 1.5–2.5 cm across. Calyx wine-red or purplish-black, campanulate, with oblong lobes often wavy-toothed at apex. Corolla saucer-shaped, violet- or purplish-blue, red at base; tube short, farinose within; lobes notched and often toothed. A form with pale yellow calyx and pure white corolla is not uncommon.

E. Nepal to S.E. Tibet. 3800–4300 m. Jun.–Aug. **Pl. 66.**

MYRSINACEAE

316 **Ardisia macrocarpa** Wallich

Forest undershrub to 1 m. Leaves narrow-lanceolate, acute at both ends, margins crisped and rounded-toothed, hairless. Flowers pink, on short stalks, in terminal umbels. Calyx with 5 narrow-oblong lobes. Corolla-lobes lanceolate acute, *c.* 5 mm. Berries bright glossy red, globose, depressed at apex.

Uttar Pradesh to Arunachal Pradesh. 1500–2400 m. May–Jul. **Pl. 68.**

317 **Ardisia solanacea** Roxb.

Evergreen undershrub to 3 m. Leaves obovate-oblong or elliptic, acute or long-pointed, fleshy, leathery, hairless. Flowers waxy pink, 15–20 mm across, in axillary clusters. Calyx-lobes rounded-ovate. Corolla-lobes 5, ovate-elliptic acute, twisted to right in bud. Fruit globose, depressed at apex, juicy, black when ripe.

Uttar Pradesh to Sikkim. To 1100 m. (India, China, S.E. Asia) Mar.–Dec. **Pl. 68.**

318 **Maesa macrophylla** See F.H. p. 253. **Pl. 68.**

319 **Myrsine semiserrata** See F.H. p. 254, p. 485. **Pl. 68.**

SAPOTACEAE

320 **Aesandra butyracea** (Roxb.) Baehni (*Bassia b.* Roxb.)

Tree to 22 m, with milky sap. Leaves crowded near ends of branches, obovate-oblong blunt, base rhomboid, woolly-haired when young, hairless when mature. Flowers white, 1.25–2.5 cm across, on more or less woolly flower-stalks, in clusters crowded amongst the leaves. Calyx-lobes 5, overlapping. Corolla-tube campanulate, with 8–10 spreading lobes, and numerous stamens. Berry ellipsoid, green, shining. Often planted,

but also indigenous. The seeds are pressed for oil.
Uttar Pradesh to Arunachal Pradesh. To 1500 m. Sep.–Dec. Pl. 69.

STYRACACEAE

321 **Styrax hookeri** C.B. Clarke
Tree to 10 m with stellately hairy young branchlets. Leaves elliptic long-pointed, regularly toothed, with small bristles on nerves above and scattered stellate hairs beneath. Flowers white, to 2.5 cm long, in lax axillary and terminal clusters. Flower-stems and calyx red-woolly. Calyx campanulate, shallowly 5–lobed. Corolla-tube short, hairy outside, with elliptic-oblong lobes. Fruit ellipsoid, hairy.
E. Nepal to Bhutan. 1800–2400 m. Jun. Pl. 68.

SYMPLOCACEAE

322 **Symplocos dryophila** C.B. Clarke
Shrub or small tree to 6 m, with hairless branchlets. Leaves narrowly obovate-lanceolate, apex fine-pointed, leathery, densely hairy beneath when young, soon becoming hairless; nerves many, midrib impressed above. Flowers yellowish-white, *c.* 4 mm, in long lax undivided sparsely tawny-haired axillary clusters. Bracts leafy; outer bracts orbicular, hairless; inner bracts oblong, hairy. Calyx lobes small, ovate. Petals 5, spreading. Disk hairless. Stamens many. Fruit globose, smooth, crowned with calyx-rim.
E. Nepal to S.W. China. 2000–3000 m. (S.E. Asia) Apr.–May. Pl. 69.

323 **Symplocos glomerata** King ex C.B. Clarke
Shrub or small tree to 5 m. Leaves lanceolate long-pointed, gland-toothed on margins, leathery, midrib impressed above, nerves conspicuous beneath; leaf-stalk glandular. Flowers yellowish-white, *c.* 4 mm, in dense axillary clusters. Bracts and bracteoles ovate, somewhat woolly-haired. Calyx rusty-woolly within, with broad rounded lobes. Corolla twice as long as calyx, with 5 spreading petals. Disk hairless. Stamens 25. Fruit cylindric, smooth, crowned by calyx-rim.
E. Nepal to S.W. China. 1800–2500 m. (S.E. Asia) Apr.–May. Pl. 69.

324 **Symplocos pyrifolia** Wallich ex G. Don
Shrub or tree to 10 m. Leaves narrowly elliptic long-pointed, base narrowed, margins entire or glandular-toothed. Flowers yellowish-white, in axillary clusters. Calyx with ovate-triangular sparsely silky-haired lobes. Petals 5, *c.* 5 mm long. Disk hairless. Stamens many. Fruit ellipsoid-cylindric, crowned by calyx-rim.
C. Nepal to Arunachal Pradesh. 1000–2000 m. Sep.–Nov. Pl. 70.

325 **Symplocos ramosissima** See F.H. p. 255. **Pl. 69**.

326 **Symplocos sumuntia** See F.H. p. 255. **Pl. 70**.

327 **Symplocos theifolia** See F.H. p. 254, p. 485. **Pl. 70**.

OLEACEAE Olive Family

328 **Jasminum dispermum** See F.H. p. 257. **Pl. 71**.

329 **Jasminum mesneyi** Hance (*J. primulinum* Hemsley ex Baker)
Shrubby rambler with angled branches, usually pendent from banks and walls. Leaves trifoliate, stalked; leaflets oblong-lanceolate, with fine-pointed apex, paler beneath. Flowers mostly double, 3.5–4.5 cm across, bright yellow with deeper golden eye, solitary on long axillary stalks. Bracts oblong acute. Calyx-lobes linear-oblong long-pointed. Corolla-tube narrowly funnel-shaped, *c.* 1 cm; lobes elliptic-rounded, spreading.
  Native of China. Commonly grown in gardens, occurring in hedges in Nepal valley. 1300–1500 m. Feb.–Mar. **Pl. 71**.

330 **Jasminum multiflorum** CHINESE JASMINE See F.H. p. 257. **Pl. 71**.

331 **Ligustrum indicum** See F.H. p. 258. **Pl. 70**.

332 **Syringa emodi** See F.H. p. 258, p. 486. **Pl. 70**.

APOCYNACEAE Oleander Family

333 **Alstonia neriifolia** D. Don
Shrub to 3 m, with milky sap and softly hairy branchlets. Leaves whorled, narrowly lanceolate long-pointed, leathery, with close parallel veins, softly hairy beneath; awl-shaped glands prominent between leaf-stalks. Flowers white, 2–2.5 cm across, in branching terminal clusters. Calyx small, with 5 triangular-ovate teeth. Corolla salver-shaped; tube cylindric; lobes oblong blunt, overlapping to left. Fruit a pair of parallel linear beaked follicles.
  C. Nepal to Bhutan. 500–900 m. Apr.–Sep. **Pl. 73**.

334 **Chonemorpha fragrans** (Moon). Alston (*C. macrophylla* G. Don)
Large climber, with stem to 25 m long and to 7.5 cm in diameter, exuding milky juice. Twigs covered with raised lenticels. Leaves large, elliptic-ovate or orbicular, abruptly pointed at apex, paler and with prominent nerves beneath. Flowers white with yellow centre, fragrant, to 7.5 cm across, in large stalked clusters. Calyx tubular, 5–lobed, with a ring of glands at base inside. Corolla salver-shaped, with obliquely wedge-shaped lobes twisted to left in bud. Fruit of 2 large straight parallel cylindric follicles.

Uttar Pradesh to Arunachal Pradesh. 400–1800 m. (India, Burma, S.E. Asia) May–Jun. Pl. 72.

335 **Plumeria rubra** FRANGI-PANI, TEMPLE-TREE See F.H. p. 260. Pl. 73.

336 **Wrightia arborea** (Dennst.) Mabberley (*W. tomentosa* Roemer & Schultes)
Deciduous shrub or tree to 10 m, with white milky juice, and with twigs woolly-hairy when young. Leaves elliptic long-pointed, woolly-hairy on both sides, margins wavy, lateral nerves prominent; leaf-stalk with gland in axil. Flowers white fading to yellow, 2.5 cm across, in erect branched terminal clusters. Flower-stalks woolly-hairy. Bracts ovate. Calyx small, woolly-hairy, with short rounded lobes. Corolla salver-shaped; tube twice as long as calyx, with corona of 5–10 orange scales in throat; lobes oblong rounded. Follicles long, pendulous, cylindric, beaked, rough with white tubercles.
Kashmir to Bhutan. To 1000 m. (India, China, S.E. Asia) May–Jul. Pl. 73.

ASCLEPIADACEAE Milkweed Family

337 **Asclepias curassavica** L.
Erect shrubby perennial weed to 1 m. Leaves opposite, oblong-lanceolate acute, narrowed to short leaf-stalk. Flowers orange-red, *c.* 5 mm, in many-flowered umbels 2–4 cm across. Sepals glandular within. Corolla wheel-shaped, with reflexed lobes. Coronal scales 5, erect, spoon-shaped. Follicles long, smooth, swollen, pointed.
Native of Tropical America. Naturalized throughout Himalaya. To 1500 m. Flowers throughout the year. Pl. 73.

338 **Ceropegia hookeri** See F.H. p. 264. Pl. 72.

339 **Ceropegia longifolia** See F.H. p. 263. Pl. 73.

340 **Ceropegia pubescens** See F.H. p. 263, p. 486. Pl. 72.

341 **Ceropegia wallichii** Wight
Perennial with stout erect stem to 50 cm, and tuberous rootstock. Leaves ovate, blunt or acute, entire, hairy on margins and on nerves beneath. Flowers purple-brown, or greenish-yellow with dark striations, in axillary clusters. Sepals small, thread-like. Corolla *c.* 3.8 cm long, tubular, with inflated base; mouth funnel-shaped, 5–angled; corolla-lobes cohering at tips, with fringe of hairs within. Follicles long, slender, tapering.
Himachal Pradesh to C. Nepal. 2500–2900 m. May–Jun. Pl. 72.

342 **Cynanchum auriculatum** See F.H. p. 261. Pl. 71.

343 **Pentasacme wallichii** Wight
Tufted hairless perennial usually growing on damp rocks beneath water-
falls, with wiry roots and pendulous stems to 80 cm. Leaves opposite,
elliptic-lanceolate long-pointed, base acute. Flowers white, c. 11 mm
long, in axillary umbels. Calyx small, with 5 elliptic acute lobes. Corolla-
tube short-campanulate, with long linear lobes gradually narrowing towards tip. Follicles paired, straight, smooth, slender.
   Uttar Pradesh to E. Nepal. 800–2400 m. Apr.–Oct. Pl. 72.

344 **Raphistemma pulchellum** Wallich
Large climber with smooth slender branches and milky juice. Leaves
ovate-cordate long-pointed, stalked. Flowers white c. 3 cm long, in
few-flowered flat-topped axillary cluster. Calyx with ovate lobes, glandular within. Corolla campanulate, 5–lobed, the lobes spreading and overlapping to right. Corona of 5 large scales with slender strap-shaped
extensions. Fruit a large solitary swollen capsule, with slightly curved
beak.
   C. Nepal to Sikkim. S.W. China. 800–1400 m. (S.E. Asia) Aug.–Sep.
Pl. 73

345 **Vincetoxicum hirundinaria** See F.H. p. 263, p. 486. Pl. 71.

LOGANIACEAE   Buddleia Family

346 **Buddleja asiatica** See F.H. p. 265. Pl. 75.

347 **Buddleja crispa** See F.H. p. 265.
This is a very variable species. The plant illustrated is the high-altitude
form with flowers in tight heads which open when the plant is almost
leafless. Pl. 75.

GENTIANACEAE   Gentian Family

348 **Gentiana algida** See F.H. p. 269, p. 488. Pl. 75.

349 **Gentiana elwesii** C.B. Clarke
Perennial with erect stem 7–20 cm. Basal leaves elliptic acute; stem-
leaves elliptic or oblong. Flowers violet-blue above, white below, in erect
terminal head. Calyx-lobes very unequal. Corolla 2–3 cm long, contracted at mouth, 5–lobed, with folds between lobes.
   E. Nepal to S.E. Tibet. 4000–4400 m. Aug.–Oct. Pl. 74.

350 **Gentiana robusta** King ex Hook.f.
Stem 20–25 cm, robust, simple, ascending. Basal leaves narrow-lanceolate acute, to 30 cm long by 2.5 cm broad, leathery, fused at base in a
tubular sheath; uppermost leaves shorter and broader at base, in a

43

crowded involucre subtending the flowers. Flowers both axillary and in a dense terminal head. Calyx half as long as corolla, papery, split to base, with 5 bristle-tipped teeth. Corolla 3–3.5 cm long, greenish-white, not spotted, somewhat inflated in middle, with 5 short triangular lobes alternating with triangular folds.

C. Nepal. Sikkim. 3500 m. Aug.–Sep. Pl. 74.

Gentiana tibetica and *Gentiana straminea* are both very similar to 350. In C. Nepal all three species occur to the north of Annapurna and Dhaulagiri. From the fact that botanists working on the Nepal flora have named and renamed material from this area as belonging to one or other of these species it is evident that they are not easy to distinguish.

351 **Gentiana stipitata** Edgew.
Perennial with one or several ascending branches 2–5 cm long bearing solitary flower. Basal leaves in crowded rosette, ovate- or obovate-lanceolate, apex fine-pointed, margins papery. Stem-leaves often somewhat coloured; lower leaves small, broadly ovate acute; upper leaves larger, lanceolate, enclosing base of flower. Calyx-lobes broadly ovate-oval, apex fine-pointed, margins papery. Corolla 2.5–3.4 cm long, usually pale blue with dark striations; corolla-lobes broadly ovate, *c.* 4 mm long by 4–5 mm across, fine-pointed at apex, and with alternating narrow-triangular acute folds half as long as lobes.

Uttar Pradesh to W. Nepal. 3600–4500 m. Aug.–Oct. Pl. 74.

352 **Gentiana urnula** See F.H. p. 271, p. 488. Pl. 74.

353 **Swertia cuneata** See F.H. p. 267. Pl. 75.

354 **Swertia multicaulis** See F.H. p. 267. Pl. 75.

355 **Swertia nervosa** (G. Don) C.B. Clarke
Annual with erect stem to 1 m. Basal leaves absent at flowering time; stem-leaves elliptic-lanceolate, 3–nerved, narrowed to base. Flowers greenish-yellow spotted with purple, in many-flowered branched axillary and terminal clusters. Sepals oblong-linear, twice as long as corolla. Corolla-lobes 4, ovate long-pointed, 6–8 mm, with gland near base.

Himachal Pradesh to S.W. China. 700–3000 m. Aug.–Oct. Pl. 74.

BORAGINACEAE   Borage Family

356 **Arnebia guttata** See F.H. p. 278. Pl. 76.

357 **Eritrichium canum** See F.H. p. 281. Pl. 76.

358 **Eritrichium nanum** subsp.villosum
The plant described on p. 281 of F.H. as *E. nanum* is the Asian form of that species. Some botanists regard it as a subspecies of the European *E. nanum*, others give it separate specific status as *E. villosum*. Pl. 76.

359 Lindelofia anchusoides   See F.H. p. 280. Pl. 77.

360 Maharanga emodi   See F.H. p. 278, p. 489. Pl. 76.

361 Microula sikkimensis   See F.H. p. 282, p. 490. Pl. 77.

362 Pseudomertensia racemosa   See F.H. p. 282. Pl. 76.

CONVOLVULACEAE   Convolvulus Family

363 Porana racemosa   Snow Creeper   See F.H. p. 286. Pl. 77.

SOLANACEAE   Potato Family

364 Nicandra physalodes   Apple of Peru   See F.H. p. 288. Pl. 77.

365 **Solanum torvum** Swartz
Prickly shrub of wastelands, with erect stem to 3 m and stellately hairy branchlets. Leaves ovate, with lobed or sinuate margins, softly hairy above, densely stellate-hairy beneath. Flowers white, 2.5–3 cm across, in dense lateral many-flowered clusters. Calyx-lobes lanceolate, sparingly hairy. Corolla softly hairy outside, with short tube and 5 triangular spreading lobes. Berry globose, smooth, yellow when ripe.
   Native of W. Indies. Naturalised C. Nepal to Bhutan. 250–750 m. Apr.–Aug. Pl. 77.

SCROPHULARIACEAE   Figwort Family

366 Hemiphragma heterophyllum   See F.H. p. 294, p. 492. Pl. 78.

367 Lagotis cashmeriana   See F.H. p. 295. Pl. 81.

368 Lagotis kunawurensis   See F.H. p. 295, Pl. 94.
This is a very variable species, and the plant shown in F.H. differs widely from the plant shown here. Pl. 81.

369 Mazus dentatus   See F.H. p. 293. Pl. 81.

370 Oreosolen wattii   See F.H. p. 296, p. 492. Pl. 78.

371 Pedicularis cheilanthifolia   See F.H. p. 300. Pl. 80.

372 Pedicularis elwesii   See F.H. p. 302. Pl. 78.

373 Pedicularis klotzschii   See F.H. p. 299. Pl. 80.

374 Pedicularis longiflora var. tubiformis   See F.H. p. 300, Pl. 96. Pl. 79.

375 Pedicularis megalantha   See F.H. p. 302. Pl. 79.

376 Pedicularis rhinanthoides   See F.H. p. 302. Pl. 78.

45

377 **Pedicularis roylei** See F.H. p. 301. **Pl. 80.**

378 **Pedicularis siphonantha** See F.H. p. 303. **Pl. 79.**

379 **Scrophularia decomposita** Royle ex Benth. (*S. lucida* auct. non L.)
Perennial with erect obscurely 4–angled stems to 1 m. Leaves 1–2–times pinnately divided; segments spreading, very unequal, oblong-ovate or lanceolate, lobed or toothed. Flowers in few-flowered short-stalked clusters forming a narrow inflorescence. Bracts small, linear. Sepals rounded, with broad papery margins. Corolla-tube inflated, 6–8 mm broad. Corolla-lobes 5; upper 4 lobes pink or brownish-purple, 2 longer than the others; lower lobe green or white, spreading.
Pakistan to C. Nepal. 1500–4000 m. May–Aug. **Pl. 82.**

380 **Veronica himalensis** See F.H. p. 297, p. 493. **Pl. 82.**

381 **Veronica lanuginosa** See F.H. p. 297, p. 493. **Pl. 78.**

382 **Veronica laxa** See F.H. p. 297. **Pl. 82.**

383 **Veronica persica** See F.H. p. 298. **Pl. 80.**

384 **Wulfenia amherstiana** See F.H. p. 296, p. 492. **Pl. 81.**

OROBANCHACEAE   Broomrape Family

385 **Aeginetia indica** See F.H. p. 304, p. 493. **Pl. 82.**

386 **Orobanche aegyptiaca** Pers. (*O. indica* Buch.–Ham. ex Roxb.)
Leafless parasite common in mustard-fields, with erect simple or branched stem 20–50 cm. Flowers violet, with white throat, 2–3 cm long, in lax terminal spike to 25 cm long. Bracts ovate, half as long as corolla-tube. Bracteoles thread-like. Calyx 4–5–toothed. Corolla downy, with curved rather slender 2–lipped tube; upper lip erect, 2–lobed; lower lip with 3 spreading lobes.
Pakistan to E. Nepal. 150–3100 m. (India, W. Asia, Europe, N. Africa). At lower altitudes Dec.–Mar., at higher altitudes Jun. **Pl. 82.**

387 **Orobanche solmsii** C.B. Clarke ex Hook.f.
Robust erect biennial or perennial parasite, with unbranched yellowish-brown stem to 50 cm or more; stem leafless, but clothed with lanceolate scales. Flowers yellowish-brown, in dense spike 10–25 cm. Bracts narrow-lanceolate, usually as long as flowers. Calyx half as long as corolla, with 2 segments, each cut to middle into 2 lanceolate lobes. Corolla 14–18 mm, curved, 2–lipped, with rounded–toothed lobes.
Pakistan to C. Nepal. 2400–3300 m. (C. Asia) Jun.–Aug. **Pl. 82.**

GESNERIACEAE   Gloxinia Family

388 Aeschynanthus hookeri C.B. Clarke
Epiphytic shrub to 1 m. Leaves fleshy, lanceolate long-pointed, entire, base wedge-shaped, midrib broad beneath. Flowers scarlet or orange, 2.5 cm long or more, in terminal umbel. Bracts narrowly oblong. Calyx scarlet, tubular, shortly and bluntly 5–lobed. Corolla tubular, unequally inflated, curved, hairy outside, 2–lipped, the lobes black-spotted. Stamens long-exserted; filaments glandular-hairy; anthers narrowly oblong. Capsule long, linear.
E. Nepal to S.W. China. 1600–2700 m. (Burma) May–Jun. **Pl. 83.**

389 Chirita bifolia D. Don
Stem erect, to 25 cm, leafless at base at flowering time, with 2 unequal leaves above. Leaves orbicular-oval, base cordate, toothed, thinly hairy on both surfaces, stalkless. Flowering stems 1–2, rarely more, rising between the 2 leaves, each stem bearing 1–3 nodding flowers c. 5 cm long. Bracts oblong, hairy. Calyx deeply 5–lobed, hairy. Corolla-tube funnel-shaped, swollen on one side, white, with golden markings within; corolla-lobes 5, rounded, blue or purple. Capsule linear, curved.
Himachal Pradesh to Bhutan. 1500–2400 m. Jun.–Aug. **Pl. 84.**

390 Chirita pumila   See F.H. p. 308. **Pl. 83.**

391 Corallodiscus lanuginosus   See F.H. p. 308, p. 494. **Pl. 84.**

392 Didymocarpus aromaticus   See F.H. p. 309. **Pl. 83.**

393 Didymocarpus oblongus Wallich ex D. Don
Stem erect, to 25 cm, softly hairy, usually leafless at flowering time except for 4 whorled leaves at apex. Leaves oblong-elliptic, coarsely toothed or lobed, softly hairy above, hairless beneath except on nerves; leaf-stalk hairy. Flowers reddish-purple, on glandular flowering stems, in branched terminal clusters. Bracts purplish, broad-ovate, fused below. Calyx purplish, campanulate, with small rounded lobes. Corolla tubular, to 1 cm, with oblique mouth and 5 rounded spreading lobes.
Himachal Pradesh to Arunachal Pradesh. 1000–3000 m. Jun.–Jul. **Pl. 84.**

394 Didymocarpus pedicellatus R. Br.
A plant of damp rocks, usually stemless, with 2 large long-stalked basal leaves. Leaves roundly ovate, unequal at base, rounded-toothed, glandular above, paler and with prominent nerves beneath. Flowers many, deep purple, to 2.5 cm long, borne in clusters at ends of erect flowering stems to 20 cm. Bracts ovate, often fused below. Calyx coloured, funnel-shaped, with rounded lobes. Corolla-tube cylindric; mouth oblique, with 5 rounded lobes. Capsule linear, beaked.
Himachal Pradesh to Arunachal Pradesh. 500–2500 m. Jul.–Sep. **Pl. 83.**

395 **Lysionotus serratus** D. Don

Perennial, epiphytic or on damp rocks, with branched stem to 50 cm. Leaves often clustered in threes, narrow-lanceolate or elliptic, long-pointed, margins toothed or wavy, with prominent oblique nerves, stalked. Flowers mauvish- or purplish-white, with yellow throat, in lax drooping clusters. Bracts ovate. Sepals narrow-lanceolate, fused at base. Corolla tubular, straight, to 2.5 cm long, inflated in middle, 2–lipped; lower lip with 2 folds within. Capsule linear.

Uttar Pradesh to Arunachal Pradesh. 1000–2400 m. (Burma) Jul.–Sep. Pl. **84.**

BIGNONIACEAE Bignonia Family

396 **Incarvillea emodi** See F.H. p. 310. Pl. **85.**

397 **Incarvillea younghusbandii** See F.H. p. 310. Pl. **85.**

ACANTHACEAE Acanthus Family

398 **Aechmanthera gossypina** See F.H. p. 312, p. 494. Pl. **86.**

399 **Echinacanthus attenuatus** (Wallich ex Nees) Nees

Herbaceous perennial with erect stem to 70 cm, softly hairy upwards. Leaves oblong-ovate, more or less toothed, stalked. Flowers mauve, c. 3 cm long, in sticky-hairy clusters. Bracts small, linear. Calyx with 5 long linear lobes. Corolla tubular, unequally swollen, with 5 rounded lobes. Capsule oblong.

C. Nepal to Sikkim. To 1200 m. Oct.–Mar. Pl. **85.**

400 **Eranthemum pulchellum** See F.H. p. 314, p. 494. Pl. **85.**

401 **Goldfussia nutans** Nees (*Strobilanthes n.* (Nees) T. Anders.)

Herbaceous perennial to 1 m, with hairy branches. Leaves ovate or broadly elliptic, long-pointed, toothed, hairy above and beneath, stalked. Flowers white, 2.5–3 cm long, in dense hairless spikes, on diverging deflexed hairy flowering-stems. Bracts large, elliptic acute, concave. Bracteoles oblong. Calyx divided nearly to base into hairless linear lobes. Corolla tubular, curved, unequally swollen, with contracted base, 5–lobed.

C. Nepal. 2100–2700 m. (Assam) Aug.–Sep. Pl. **86.**

402 **Strobilanthes atropurpureus** Nees

Forest perennial with erect 4–grooved stem, to 50 cm but often much smaller, hairy upwards. Leaves ovate-lanceolate long-pointed, base narrowed to stalk, coarsely toothed, white-hairy, with 6 pairs of lateral nerves. Flowers blue, c. 4 cm long, solitary or paired, forming a leafy interrupted spike. Bracts persistent, leaf-like. Sepals linear blunt, with

spreading hairs. Corolla-tube broad, curved, with oblique mouth and
short rounded lobes.
Pakistan to C. Nepal. 1300–3600 m. Jun.–Aug. **Pl. 86.**

403 **Thunbergia coccinea** See F.H. p. 314. **Pl. 86.**

404 **Thunbergia fragrans** See F.H. p. 313. **Pl. 86.**

VERBENACEAE Verbena Family

405 **Callicarpa arborea** Roxb.
Evergreen tree to 10 m with stellately hairy branchlets. Leaves opposite,
ovate-lanceolate or narrowly oblong, long-pointed, margins undulate,
hairy on both sides when young, densely felted beneath when mature,
stalked. Flowers mauve, *c.* 3 mm, in many-flowered stellately hairy
branched axillary clusters. Calyx small. Corolla tubular-campanulate,
with 4 lobes. Stamens exserted. Fruit purplish-black, globose.
Uttar Pradesh to Bhutan. 250–2000 m. (India, China, S.E. Asia)
Apr.–May. **Pl. 87.**

406 **Callicarpa macrophylla** Vahl
Evergreen shrub to 3 m. Twigs, leaf-stalks, flower-stalks and undersur-
faces of leaves densely woolly-hairy. Leaves opposite, ovate-oblong-lanceolate, long-pointed, rounded-toothed, stalked. Flowers mauvish-pink,
in rounded many-flowered branching axillary clusters: Calyx campanulate, minutely 4–lobed. Corolla 2.5 mm long, with cylindric tube and 4
spreading lobes. Stamens exserted. Fruit white, globose, succulent.
Kashmir to Bhutan. To 1500 m. (India, China, S.E. Asia) Jun.–Sep.
Pl. 87.

407 **Caryopteris odorata** See F.H. p. 318, p. 495. **Pl. 84.**

408 **Clerodendrum philippinum** Schauer
Shrub to 4 m. Leaves opposite, broadly ovate acute, margins undulate or
coarsely toothed, softly hairy above and beneath; leaf-stalk with a few
large glands near top. Flowers white tinged with pink, fragrant, 2.5 cm
across, in compact terminal cluster to 10 cm across. Calyx red, with 5
lanceolate long-pointed lobes. Corolla doubled or trebled, Stamens 4,
long-exserted.
Native of China. Grown in Himalaya as a medicinal plant, and semi-
naturalized around villages. To 2000 m. Aug.–Dec. **Pl. 87.**

409 **Clerodendrum serratum** See F.H. p. 316. **Pl. 87.**

410 **Clerodendrum viscosum** See F.H. p. 317, p. 495. **Pl. 88.**

411 **Duranta repens** See F.H. p. 318. **Pl. 88.**

412 **Holmskioldia sanguinea** See F.H. p. 318, p. 495. **Pl. 84.**

413 **Vitex negundo** See F.H. p. 316, p. 495. **Pl. 88.**

49

LABIATAE   Mint Family

414   **Ajuga lobata**   See F.H. p. 320, p. 496. Pl. 89.

415   **Ajuga lupulina**   See F.H. p. 319.
The large densely overlapping prominently net-veined bracts, which almost conceal the flowers, enlarge and turn red after flowering. Pl. 89.

416   **Anisomeles indica** (L.) O. Kuntze (*A. ovata* R. Br.)
Woolly-haired annual weed of wastelands and cultivated areas, with acutely angled erect branching stem to 150 cm. Leaves ovate long-pointed, irregularly toothed, stalked. Flowers mauve or purple, 1.25 cm long, in numerous dense whorls. Calyx hairy, with 5 triangular-lanceolate long-pointed lobes. Corolla-tube short, 2–lipped; upper lip entire; lower lip broad, spreading, with mid-lobe larger and notched at apex. Stamens 4, in unequal pairs.
Uttar Pradesh to Sikkim. 200–2400 m. (India, China, S.E. Asia) Aug.–Nov. Pl. 90.

417   **Colebrookea oppositifolia** Smith
Shrub to 5 m, with densely silky-hairy twigs, leaf-stalks and inflorescences. Leaves opposite, elliptic-oblong long-pointed, rounded-toothed, base acute, softly hairy beneath. Flowers white, 2.5 mm long, in whorls forming dense axillary and terminal branched spike-like clusters. Bracts solitary, linear. Bracteoles whorled, fused at base. Calyx white-hairy, with 5 awl-shaped teeth. Corolla-tube very short, with 4 lobes. Fruit a minute nutlet, arranged in catkin-like spikes which are conspicuous for most of the year. Often occurring gregariously. The plant has medicinal uses.
Pakistan to S.W. China. To 1700 m. (India, S.E. Asia) Dec.–Apr. Pl. 92.

418   **Coleus barbatus**   See F.H. p. 321, p. 496. Pl. 90.

419   **Dracocephalum heterophyllum**   See F.H. p. 333. Pl. 90.

420   **Glechoma nivalis**   See F.H. p. 332, p. 497. Pl. 92.

421   **Glechoma tibetica** Jacquem.
Dwarf stemless softly hairy high-altitude perennial of Tibetan borderlands, with long branching roots penetrating deeply into screes. Leaves in a crowded rosette, fan-shaped and with a wedge-shaped base, margins rounded-toothed and sometimes pinkish, wrinkled, stalked. Flowers white tinged with pink, *c.* 18 mm long, in few-flowered axillary clusters. Calyx oblique, 3–toothed, woolly. Corolla-tube straight, funnel-shaped, with small 2–lipped limb. Nutlets linear-oblong, smooth.
Kashmir to Himachal Pradesh. 4500–5300 m. (W. Tibet) Jul.–Aug. Pl. 89.

422   **Mentha longifolia**   Horse-Mint   See F.H. p. 336, p. 498. Pl. 89.

423   **Nepeta coerulescens**   See F.H. p. 329. Pl. 91.

424   **Nepeta floccosa**   See F.H. p. 331. **Pl. 91.**

425   **Nepeta lamiopsis**   See F.H. p. 330. **Pl. 91.**

426   **Nepeta longibracteata**   See F.H. p. 330, p. 497. **Pl. 91.**

427   **Nepeta nervosa**   See F.H. p. 329. **Pl. 92.**

428   **Pogostemon benghalensis** (Burm.f.) Kuntze (*P. plectranthoides* Desf.)
Shrubby erect grey-hairy aromatic perennial to 2.5 m, usually growing
gregariously and often in association with *Clerodendrum viscosum*.
Branches purple, rounded. Leaves opposite, ovate acute, double-toothed,
stalked. Flowers 8 mm long, whorled, in dense spikes forming a pyramidal
terminal inflorescence. Bracts ovate acute, glandular. Calyx pinkish-
mauve, hairy, with short triangular-lanceolate hair-fringed teeth. Corolla
white, more or less 2–lipped, 4–lobed, the lower lobe largest. Stamens 4,
exserted, with purple-bearded filaments. Nutlets minute, broadly ellipsoid, shining, dark brown when ripe.
   Himachal Pradesh to E. Nepal. To 1300 m. (India) Dec.–May. **Pl. 92.**

429   **Stachys melissaefolia**   See F.H. p. 327. **Pl. 91.**

430   **Thymus linearis**   See F.H. p. 335, p. 498. **Pl. 92.**

AMARANTHACEAE   Cockscomb Family

431   **Cyathula capitata** Moq.
Resembling 432, but more slender and herbaceous, and only sparsely
hairy. Flowers with hooked bristles, in glistening heads 2.5–3 cm across;
heads few or solitary, not arranged in spikes.
   Kashmir to Bhutan. 1300–2900 m. (S.E. Asia) Jul. Fruiting Sep.–Nov.
Pl. 93.

432   **Cyathula tomentosa** (Roth) Moq.
Deciduous shrub to 3 m. Stems erect or decumbent, woody at base.
Branches woolly-hairy, swollen at nodes. Leaves elliptic, acute or long-
pointed, base narrowed, adpressed-hairy above, silky-hairy beneath,
short-stalked. Flowers greenish-white, in dense globose heads 2.5–4 cm
across. Flower-heads forming axillary and terminal spikes, the basal part
of the spike often interrupted. Bracts ovate, with fine-pointed apex.
Perfect flowers 1–2 in each cluster surrounded by imperfect flowers which
are reduced to a single sepal with rigid hooked bristle-tip. Fruit enclosed
in the perianth, with enlarged hooked tips.
   Himachal Pradesh to S.W. China. 1400–2400 m. Jul.–Aug. Fruiting
Sep.–Dec. **Pl. 93.**

CHENOPODIACEAE   Goosefoot Family

433   **Chenopodium foliolosum**   See F.H. p. 340. **Pl. 94.**

POLYGONACEAE Dock and Rhubarb Family

434 **Aconogonum alpinum** See F.H. p. 343. **Pl. 94.**

435 **Aconogonum campanulatum** See F.H. p. 343.
The plant shown on Pl. 106 in F.H. was named erroneously in the first printing as this species. It is in fact a photo of *Aconogonum molle*. **Pl. 96.**

436 **Aconogonum molle** See F.H. p. 342, Pl. 106. **Pl. 96.**

437 **Aconogonum tortuosum** See F.H. p. 342, p. 499. **Pl. 94.**

438 **Bistorta affinis** See F.H. p. 343, Pl. 106. **Pl. 95.**

439 **Bistorta amplexicaulis** See F.H. p. 344, p. 499, Pl. 106.
The photo shown in F.H. is of the variety found in C. & E. Nepal with pendulous flower-spikes. The plant illustrated here is the more usual and widespread form with erect spikes. **Pl. 96.**

440 **Bistorta emodi** See F.H. p. 344. **Pl. 96.**

441 **Bistorta millettii** See F.H. p. 345. **Pl. 93.**

442 **Bistorta vaccinifolia** See F.H. p. 344, p. 500. **Pl. 95.**

443 **Fagopyrum esculentum** Buckwheat See F.H. p. 347. **Pl. 95.**

444 **Persicaria polystachya** See F.H. p. 346. **Pl. 97.**

445 **Polygonum paronychioides** C. Meyer ex Hohen.
Perennial of drier areas, with woody rootstock and spreading much-branched stems 2.5–10 cm, forming loose mats. Leaves fleshy, linear, with fine-pointed apex and recurved margins. Stipules large, whitish, papery, on young shoots concealing both leaves and stem. Flowers pink, c. 3 mm across, stalkless, in axillary clusters surrounded by the leaves and stipules.
Afghanistan to Himachal Pradesh. 2700–4300 m. Jun.–Aug. **Pl. 93.**

446 **Rheum acuminatum** See F.H. p. 348. **Pl. 97.**

447 **Rheum webbianum** See F.H. p. 348. **Pl. 98.**

448 **Rumex hastatus** See F.H. p. 349, p. 500. **Pl. 97.**

449 **Rumex nepalensis** See F.H. p. 349, p. 500. **Pl. 98.**

450 **Rumex patentia** L. subsp. **tibeticus** (Rech.f.) Rech.f.
Perennial with stout erect stem to 120 cm, branched at top. Lower leaves oblong-lanceolate acute, margins often wavy, with long leaf-stalk channelled above; upper leaves smaller, lanceolate. Flowers small, yellowish-green, in whorls arranged in numerous spike-like clusters forming a long compact leafy inflorescence. Fruit enclosed by rounded net-veined valves 4.5–6 mm long and 4–5 mm wide, with notched or cut-off base and more or less acute apex.

Afghanistan to Himachal Pradesh. Common in cultivated areas of Ladakh and Lahul. 2100–4100 m. (C. Asia, Iran) Jun.–Jul. Pl. 98.

PIPERACEAE Pepper Family

451 **Peperomia tetraphylla** (Forst.f.) Hook. & Arn. (*P. reflexa* (L.f.) A. Dietrich)
Stems leafy, 7.5–25 cm, rooting at nodes on mossy trees and rocks. Leaves usually in whorls of 4, fleshy, orbicular-elliptic. Flowers green, minute, in stout erect axillary and terminal spikes 12–37 mm. Perianth nil. Stamens 2.
Kashmir to Bhutan. 1000–2500 m. (India, China, S.E. Asia, Africa, America) Mar.–Nov. Pl. 93.

452 **Piper boehmeriaefolia** (Miq.) DC.
Hairless climber on rocks and trees. Leaves obliquely oblong long-pointed, base very unequal, with 5–7 distant nerves, shortly stalked. Flowers pale green, minute, in pendent flexuous unisexual spikes 7.5–15 cm long; spikes solitary on slender stalks. Fruit a globose berry 3 mm across, borne in dense pendent cylindric spikes.
C. Nepal to Bhutan. To 1500 m. (India, Burma) Apr.–May. Pl. 99.

LAURACEAE Laurel Family

453 **Actinodaphne obovata** (Nees) Blume
Evergreen tree to 15 m, with greyish bark peeling off in thin papery flakes; young parts rusty-hairy. Leaves in whorls, obovate or elliptic-oblong, acute or blunt, 3–nerved from base, leathery, shining above, whitish beneath, stalked. Flowers pale yellow. Male flowers *c*. 1.25 cm across, with short perianth-tube and papery segments. Female flowers smaller, in loose branched clusters. Fruit ellipsoid, seated on perianth-tube.
E. Nepal to Bhutan. 450 m. (Burma, S.E. Asia) Mar.–May. Pl. 99.

454 **Cinnamomum tamala** See F.H. p. 351, p. 501. Pl. 99.

455 **Lindera neesiana** (Wallich ex Nees) Kurz
Deciduous tree to 8 m. Leaves aromatic when crushed, ovate-oblong-lanceolate, usually long-pointed, entire, 3–nerved from base. Flowers yellow, in few-flowered globose heads *c*. 8 mm across, the heads solitary or clustered. Bracts 4, papery. Perianth-segments 6, rounded. Fruit globose, seated on perianth-tube.
C. Nepal to Bhutan. 1800–2700 m. (Burma) Oct.–Nov. Pl. 100.

456 **Litsea doshia** See F.H. p. 353, p. 501. Pl. 100.

PROTEACEAE — LORANTHACEAE

457 **Neolitsea pallens** See F.H. p. 353, p. 501. Pl. 101.

458 **Persea odoratissima** See F.H. p. 352. Pl. 100.

459 **Phoebe lanceolata** (Nees) Nees
Evergreen tree to 15 m. Leaves oblong-lanceolate long-pointed, narrowed to base, blue-green and with prominent lateral nerves beneath. Flowers greenish-yellow, in lax hairless stalked clusters. Perianth cup-shaped, c. 2.5 mm long; perianth-lobes hairless outside, downy within, enlarged and hardened in fruit. Fruit black, ellipsoid, seated on enlarged perianth.
Uttar Pradesh to Bhutan. To 1100 m. (India, Burma) Mar.–May. Pl. 99.

PROTEACEAE

460 **Grevillea robusta** Cunn. ex R.Br. SILKY OAK
Evergreen tree to 25 m or more. Leaves pinnate; leaflets entire or deeply pinnate-lobed; ultimate segments linear-oblong acute, margins recurved, dark green above, grey-silky beneath. Flowers orange, in clusters to 10 cm long, from old wood of dwarf leafless branches. Calyx absent. Corolla-tube 1.25 cm long, rolled back, splitting down one side; limb ovoid, with 4 segments which cohere long after tube has opened. Fruit obliquely oblong, compressed, tipped with persistent style.
Native of Australia. Often planted along roads in Himalaya, especially in Nepal valley. To 1500 m. Apr.–May. Pl. 104.

ELAEAGNACEAE Oleaster Family

461 **Hippophae rhamnoides** subsp. **turkestanica** See F.H. p. 357. Pl. 101.

462 **Hippophae salicifolia** See F.H. p. 357. Pl. 101.

LORANTHACEAE Mistletoe Family

463 **Dendropthoe falcata** (L.f.) Etting. (*Loranthus longiflorus* Desr.)
Common parasite on subtropical trees. Leaves leathery, stalked, with red midrib, very variable in shape, ovate obovate or elliptic, sometimes oblique. Flowers orange or scarlet, 2.5–5 cm long, in dense 1–sided upcurved axillary clusters. Calyx cup-shaped. Corolla-tube hairless, gradually expanding towards apex, with 5 greenish-yellow spreading linear lobes shorter than tube. Berry black when ripe, crowned with calyx. The bark is used medicinally.
Uttar Pradesh to Bhutan. To 900 m. (India, S.E. Asia, Australia) Oct.–Mar. Pl. 100.

54

EUPHORBIACEAE

EUPHORBIACEAE   Spurge Family

464 **Euphorbia helioscopia** SUN SPURGE See F.H. p. 362. **Pl. 102.**

465 **Euphorbia prolifera** Buch.–Ham. ex D. Don
Hairless milky-juiced perennial with several erect stems to 60 cm, and often with barren densely leafy rooting shoots spreading from near base. Stem-leaves linear-oblong, acute or blunt, entire, stalkless; floral leaves elliptic blunt. Involucres 3 mm across, with 4–5 teeth, in umbellate clusters. Glands yellow, crescent-shaped, usually with short blunt horns. Capsule globose, smooth.
Pakistan to S.W. China. 900–1400 m. Apr.–May. **Pl. 102.**

466 **Euphorbia pulcherrima** POINSETTIA See F.H. p. 360. **Pl. 102.**

467 **Euphorbia sikkimensis** Boiss.
Perennial with woody rootstock, and with stout stem to 120 cm, branched above. Leaves linear-oblong acute, narrowed to short leaf-stalk; floral leaves whorled; 3–4 involucral leaves yellow, ovate-oblong blunt. Involucres hemispheric, 3 mm across, hairy within, with ovate hair-fringed lobes and transversely oblong glands. Styles very slender, shortly 2-lobed. Capsule smooth. The plant illustrated was collected in the Dudh Kosi valley of E. Nepal by Mr. A. D. Schilling. It differs in some respects from the species described above. See note on p. xi.
E. Nepal to Bhutan. 2400 m. Apr.–May. **Pl. 103.**

468 **Euphorbia stracheyi** See F.H. p. 361. **Pl. 102.**

469 **Mallotus philippinensis** (Lam.) Muell.–Arg.
Evergreen tree to 8 m, with rusty-hairy young shoots and inflorescence. Leaves alternate, closely dotted with minute red glands; leaf-blade variable, ovate-oblong, elliptic or lanceolate, fine-toothed or entire; leaf-stalk softly hairy, with small gland on each side at top. Flowers yellowish, c. 4 mm across. Male flowers in erect clustered terminal spikes 10–25 cm long. Female flowers in solitary or several terminal spikes 2.5–10 cm long. Fruit a 3-lobed capsule, covered when ripe with red resinous powder which yields an orange-yellow dye.
Pakistan to Bhutan. To 1800 m. (India, China, S.E. Asia, Australia) Sep.–Nov. **Pl. 103.**

470 **Phyllanthus emblica** L.
Deciduous tree to 15 m. Leaves linear-oblong blunt, entire, hairless, almost stalkless, arranged in two ranks on slender branchlets and thus appearing to form a pinnate leaf. Flowers minute, greenish-yellow, in axillary clusters on leafy twigs, often on the leafless part below the leaves. Sepals 5–6, oblong-obovate. Petals nil. Fruit globose, pale yellow tinged with pink.
Pakistan to Bhutan. To 1400 m. (India, China, S.E. Asia) Apr.–May. **Pl. 104.**

55

471 **Sapium insigne** (Royle) Benth. ex Hook.f.
Deciduous tree to 10 m, with furrowed bark and milky juice. Leaves crowded towards ends of branches, elliptic-lanceolate-oblong, acute or long-pointed, rounded-toothed, glossy; leaf-stalk with 2 green glands. Flowers appearing before the leaves, very small, yellowish- or brownish-green, in stout erect terminal spikes 7.5–20 cm long. Fruit very prominent when tree is leafless, red-brown, glossy, ovoid, in erect spikes.
Kashmir to Bhutan. 500–1800 m. (India, S.E. Asia) Nov.–Mar. **Pl. 103.**

472 **Trewia nudiflora** L.
Deciduous tree to 12 m. Leaves opposite, ovate long-pointed, entire, base usually cordate, hairless when mature. Male flowers yellow, 8 mm across, in numerous 1–4–flowered clusters forming a long lax drooping inflorescence 10–23 cm long, appearing with or before the new leaves. Female flowers green, solitary or 2–3 together, on axillary stalks. Fruit green, depressed globose, somewhat succulent.
Uttar Pradesh to E. Nepal. To 1800 m. (India, China, S.E. Asia) Feb.–Mar. **Pl. 101.**

BUXACEAE   Box Family

473 **Sarcococca coriacea**   See F.H. p. 363. **Pl. 101.**

474 **Sarcococca wallichii** Stapf
Hairless evergreen undershrub to 1½ m. Leaves lanceolate long-pointed, bright green, leathery, midrib prominent, stalked. Flowers in axillary clusters; upper clusters male, white; lower clusters female, green. Sepals of male flowers ovate-elliptic blunt, minutely fringed; stamens white, exserted. Sepals of female flowers ovate acute, fringed. Close to *Sarcococca saligna* of W. Himalaya (See F.H. p. 363), but sepals of male flowers bigger (to 4.5 mm long), and leaves broader (usually 2–3.5 cm across). Fruit blue-black, ovoid-globose. Seeds keeled on sides.
C. Nepal to S.W. China. 2100–2900 m. Oct. **Pl. 101.**

URTICACEAE   Nettle Family

475 **Boehmeria platyphylla**   See F.H. p. 364, p. 503. **Pl. 104.**

476 **Boehmeria rugulosa** Wedd.
Tree to 10 m. Leaves alternate, elliptic-lanceolate long-pointed, rounded-toothed, strongly 3–nerved, hairless above, white-velvety beneath, leathery, stalked. Flowers greenish-yellow, in dense unisexual clusters 5–7 mm across arranged in 1–3 simple axillary spikes 5–15 cm long. Male flowers globose, *c.* 1.7 mm across. Female flowers ovoid, *c.* 1 mm across,

white-hairy. Achenes elliptic, inflated on one side, acute at both ends. Uttar Pradesh to Bhutan. To 1700 m. Aug.–Sep. Pl. **104**.

477 **Debregeasia salicifolia**   See F.H. p. 364. Pl. **103**.

478 **Elatostema sessile** Forster & Forster f.
Stem erect, unbranched, to 60 cm, creeping at base. Leaves obliquely oblanceolate or oblong, long-pointed, coarsely toothed, base obliquely wedge-shaped, stalkless or short-stalked. Flowers white, in crowded unisexual axillary heads 3–15 mm across; heads solitary or 2–3 together, stalkless or very short-stalked. Male flowers with oblong perianth-segments 1–1.5 mm long. Female flowers bristly-haired. Achenes ellipsoid, angled.
Himachal Pradesh to Bhutan. 900–3000 m. (India, China, S.E. Asia) May–Oct. Pl. **106**.

479 **Girardinia diversifolia**   See F.H. p. 365, p. 503. Pl. **105**.

480 **Lecanthus peduncularis**   See F.H. p. 367, p. 503. Pl. **105**.

481 **Pilea glaberrima** (Blume) Blume (*P. smilacifolia* Wedd.)
Hairless perennial of damp shady places, with stem to 2 m, woody below, and with angled branchlets. Leaves obliquely elliptic-lanceolate long-pointed, entire, 3–nerved, narrowed to leaf-stalk. Stipules persistent, triangular or lanceolate. Flowers white or cream, in short-stalked branched clusters 1–3 cm long. Male flowers globose, 1–1.5 mm across. Female flowers with ovary 0.5 mm. Achenes ovoid, compressed, smooth, *c.* 1.25 mm.
C. Nepal to Bhutan. To 1700 m. (Burma, S.E. Asia) Apr.–Sep. Pl. **105**.

482 **Pilea umbrosa** Blume
Forest perennial with erect stem 30–50 cm, often densely clothed with flexuous cellular hairs. Leaves opposite, broadly elliptic-ovate or oblong, acute or long-pointed, coarsely toothed, 3–nerved, base cordate or rounded, stalked. Stipules fused at base. Flowers minute, greenish-white, pinkish or purple. Male flowers in widely branched clusters 10–15 cm long, discharging clouds of pollen when shaken. Female clusters to 6 cm long. Achenes *c.* 1 mm, smooth, straight.
Pakistan to S.W. China. 1200–2500 m. (Burma) Jun.–Aug. Pl. **105**.

MORACEAE   Mulberry or Fig Family

483 **Ficus benghalensis** BANYAN   See F.H. p. 369, p. 504. Pl. **107**.

484 **Ficus benjamina** L. var. **nuda** (Miq.) Barrett
Tree to 25 m, with drooping branches. Leaves ovate-elliptic, abruptly pointed, base rounded or acute, hairless, shining, stalked; nerves many,

close and straight, converging within margin. Figs in axillary pairs, smooth, globose, stalkless, *c.* 18 mm across, orange-yellow when ripe. Planted as a shade-tree.

W. to C. Nepal. 600–1200 m. (India, China, S.E. Asia) Figs ripen Apr. **Pl. 106.**

485 **Ficus glaberrima** Blume
Tree to 25 m, with trunks of mature trees usually buttressed at base. Leaves elliptic-oblong or ovate-lanceolate, abruptly long-pointed, entire, hairless, shining dark green above, paler beneath; lateral nerves impressed above, prominent beneath. Figs axillary, 1–2 together, obovoid, tubercled, *c.* 6 mm across, orange or red when ripe. This species often starts as an epiphyte.

C. Nepal to Bhutan. 600–1500 m. (India, China, S.E. Asia) Figs ripen during winter. **Pl. 107.**

486 **Ficus hederacea** Roxb. (*F. scandens* Roxb.)
Large evergreen climber on rocks and trees, adhering by means of rooting branches. Leaves ovate-elliptic acute, entire, leathery, with 4–5 pairs of arching lateral nerves. Figs axillary, solitary or paired, globose or ovoid, 8–10 mm across, orange-yellow or pinkish with distant pale dots when ripe.

Uttar Pradesh to Arunachal Pradesh. 500–1400 m. (India, China, S.E. Asia) Figs ripen Feb.–Apr. **Pl. 107.**

487 **Ficus hispida** L.
Shrub or small tree to 10 m, with hollow twigs and smooth grey-green bark, common around villages and often lopped for fodder. Leaves ovate or obovate-oblong, acute or blunt, usually toothed, bristly-hairy above and beneath, base wedge-shaped or slightly cordate, stalked. Stipules ovate-lanceolate. Figs ovoid, 1.25–2.5 cm across, bristly-hairy, yellow when ripe. On young plants the figs are axillary, on mature plants they are borne on leafless shoots from main stem or larger branches.

Himachal Pradesh to Bhutan. 450–1100 m. (India, China, S.E. Asia, Australia) Figs ripen Jul. **Pl. 107.**

488 **Ficus oligodon** See F.H. p. 370. **Pl. 106.**

489 **Ficus religiosa** Pipal See F.H. p. 369, p. 504. **Pl. 106.**

490 **Ficus semicordata** See F.H. p. 370. **Pl. 106.**

JUGLANDACEAE Walnut Family

491 **Engelhardia spicata** See F.H. p. 372, p. 505. **Pl. 109.**

MYRICACEAE

492 **Myrica esculenta** See F.H. p. 373, p. 505. **Pl. 108.**

BETULACEAE Birch Family

493 **Betula utilis** See F.H. p. 374, p. 506. **Pl. 108.**

SALICACEAE Willow Family

494 **Populus jacquemontiana** var. **glauca** See F.H. p. 383. **Pl. 108.**

495 **Salix babylonica** L.
Tree to 18 m with pendent branchlets hairless except near nodes, commonly planted along streams and irrigation channels. Trees in the Himalaya are usually male, in Europe usually female. Leaves lanceolate long-pointed, narrowed to base, finely toothed, silky when young, hairless later, light green above, blue-green beneath. Catkins dense, slender, almost stalkless, with 1–3 small leaves at base, appearing with the young leaves. Scales persistent, lanceolate-ovate, yellowish.
  Probably originating in N. China. Planted in Himalaya to 3650 m. Apr. **Pl. 109.**
  Salix babylonica is very prominent in the Kashmir valley, where it is much planted along water-channels. The other willow commonly planted there is generally known as *Salix alba*, although it differs considerably from the European form of that species.

EPHEDRACEAE Joint-Pine Family

496 **Ephedra gerardiana** See F.H. p. 384, p. 510. **Pl. 110.**

PINACEAE Pine Family

497 **Abies densa** E. HIMALAYAN SILVER FIR See F.H. p. 386. **Pl. 110.**

498 **Abies spectabilis** HIMALAYAN SILVER FIR See F.H. p. 386, p. 510. **Pl. 110.**

499 **Cedrus deodara** DEODAR, HIMALAYAN CEDAR See F.H. p. 386, p. 510. **Pl. 112.**

500 **Larix griffithiana** E. HIMALAYAN LARCH See F.H. p. 387. **Pl. 112.**

501 **Larix himalaica** See F.H. p. 387, p. 510. **Pl. 111.**

502 **Picea smithiana** W. HIMALAYAN SPRUCE See F.H. p. 385, p. 510. Pl. 112.

503 **Pinus roxburghii** LONG-LEAVED OR CHIR PINE See F.H. p. 388, p. 511. Pl. 113.

504 **Pinus wallichiana** HIMALAYAN BLUE PINE See F.H. p. 387, p. 511. Pl. 110.

505 **Tsuga dumosa** HIMALAYAN HEMLOCK See F.H. p. 385, p. 511. Pl. 110.

CUPRESSACEAE   Cypress Family

506 **Cupressus torulosa** W. HIMALAYAN CYPRESS See F.H. p. 389, p. 511. Pl. 111.

507 **Juniperus communis** COMMON JUNIPER See F.H. p. 389, p. 512. Pl. 112.

508 **Juniperus indica** BLACK JUNIPER See F.H. p. 390, p. 512. Pl. 113.

509 **Juniperus macropoda** PENCIL CEDAR See F.H. p. 390, p. 512.
This species is close to and perhaps conspecific with *Juniperus excelsa* of W. Asia and Greece. Pl. 111.

510 **Juniperus recurva** DROOPING JUNIPER See F.H. p. 390, p. 512. Pl. 113.

511 **Juniperus squamata** See F.H. p. 390, p. 512. Pl. 113.

PODOCARPACEAE

512 **Podocarpus neriifolius** D. Don
Evergreen tree to 15 m or more. Leaves linear long-pointed, leathery, stalkless, with midrib prominent above and beneath. Male flowers clustered in dense narrow-cylindric catkins. Female flowers solitary, axillary. Fruit globose, fleshy, 1–1.5 cm, bearing 1–2 seeds 7–8 mm long surrounded by a yellow-orange fleshy scale.
C. Nepal to Sikkim. Rare in our area, but not uncommon in Madi Khola near Pokhara, C. Nepal. 1000–1100 m. (China, Burma, S.E. Asia) Flowers during the monsoon. Pl. 111.

TAXACEAE   Yew Family

513 **Taxus baccata** subsp. **wallichiana** HIMALAYAN YEW See F.H. p. 391. Pl. 112.

ORCHIDACEAE   Orchid Family

TERRESTRIAL ORCHIDS

514   **Coeloglossum viride** (L.) Hartman
Stem erect, to 25 cm, from ovoid often lobed tuberous roots. Leaves usually 3–4, narrow-oblong, acute or blunt, many-nerved. Flowers greenish-red, in lax spike 2–10 cm. Sepals 4.5–6 mm long, ovate-oblong, concave, hooded at apex. Petals linear-oblong, erect. Lip linear-oblong, 3–toothed at apex, pale brown. Spur short, blunt.
Kashmir to S.E. Tibet. 3100–3900 m. (C. & N. Asia, Europé) Jun.–Aug. **Pl. 114.**

515   **Eulophia herbacea** Lindley
Stem 60–90 cm, from basal pseudo-bulbs. Leaves plaited, many-nerved, linear-lanceolate. Floral bracts lanceolate long-pointed, papery. Sepals green, 3.75 cm long, linear-lanceolate long-pointed, spreading. Petals white, elliptic-lanceolate. Lip white, obovate-oblong, with rounded lobes. Spur very short, rounded, somewhat curved.
Kashmir to C. Nepal. 1300–2000 m. (India) Jun.–Jul. **Pl. 114.**

516   **Galearis spathulata**   See F.H. p. 396. **Pl. 115.**

517   **Goodyera fusca**   See F.H. p. 397. **Pl. 118.**

518   **Habenaria stenopetala** Lindley
Stem erect, 30–100 cm, clothed below with closely fitting sheaths. Leaves oblong- or ovate-lanceolate, acute or long-pointed, margins undulate, sheathing or stem-clasping at base; uppermost leaves bract-like. Flowers pale green, *c.* 2.5 cm across, in crowded cluster 10–18 cm long. Sepals oblong-lanceolate long-pointed, spreading or reflexed. Petals linear, shorter than sepals. Lip reddish-brown, with 3 narrow fleshy lobes. Spur slender, compressed, upturned.
Kashmir to Sikkim. 1500–2500 m. Aug.–Oct. **Pl. 114.**

519   **Neottianthe calcicola** (W.W. Smith) Schlechter
Tuberous orchid of drier areas, with erect stem to 20 cm. Leaves linear-lanceolate, blunt or acute. Flowers pinkish-mauve, in 6–12–flowered clusters. Sepals obliquely lanceolate, *c.* 7 mm. Petals linear, as long as sepals. Lip marked with white at base, and with 3 linear lobes.
W. Nepal to S.W. China. 3700–5000 m. Aug.–Sep. **Pl. 114.**

520   **Neottianthe secundiflora** (Hook.f.) Schlechter (*Habenaria s.* Hook.f.)
Tuberous orchid of drier areas. Leaves 2–3, linear long-pointed, with tubular sheaths at base. Flowers mauvish-pink, 9 mm long, in crowded one-sided somewhat recurved spike-like clusters. Floral bracts lanceolate long-pointed. Sepals lanceolate; petals linear-sickle-shaped; sepals and petals together forming a hood. Lip deflexed, 3–lobed at apex. Spur conical, with blunt apex.
Uttar Pradesh to S.E. Tibet. 2500–3200 m. Jul.–Sep. **Pl. 114.**

ORCHIDACEAE

521 **Platanthera arcuata** Lindley (*Habenaria a.* (Lindley) Hook.f.)
Stem to 80 cm, thick and leafy. Leaves oblong-lanceolate acute, with
sheathing at base; upper leaves linear-lanceolate. Flowers white, in broad
many-flowered spike. Floral bracts leaf-like, linear-lanceolate. Dorsal
sepal hooded and beaked; lateral sepals larger, 9 mm long, oblong blunt,
deflexed. Petals small, linear, papery. Lip twice as long as sepals, linear,
with reflexed margins. Spur to 6 cm, linear acute, incurved.
Uttar Pradesh to S.E. Tibet. 2600–3100 m. Jun.–Aug. Pl. 114.

522 **Spathoglottis ixioides** See F.H. p. 401, p. 513. Pl. 115.

EPIPHYTIC ORCHIDS

523 **Aerides odorata** Lour.
Leaves to 25 cm long, leathery, flat, linear, keeled, tapering to unequally
bi-lobed apex. Flowers white spotted with pink, fragrant, *c.* 2.5 cm long
by 2 cm across, in deflexed many-flowered clusters as long as or longer
than leaves. Sepals spreading, the dorsal sepal broadly elliptic-ovate, the 2
lateral sepals somewhat larger. Petals oblong sickle-shaped, shorter than
sepals. Lip with shallow lateral lobes and linear apical lobe. Spur large,
funnel-shaped, curved forwards.
C. Nepal to Sikkim. 200–1200 m. (India, China, S.E. Asia) May–Jun.
Pl. 118.

524 **Coelogyne cristata** See F.H. p. 402. Pl. 115.

525 **Coelogyne fuscescens** Lindley
Pseudobulbs long, cylindric, furrowed, borne about 2.5 cm apart on stout
root. Leaves paired, plaited, elliptic acute, narrowed to short leaf-stalk.
Flowers pale yellowish-brown, *c.* 3.8 cm long, in drooping few-flowered
clusters rather shorter than the leaves. Sepals oblong-lanceolate acute,
concave, spreading. Petals linear acute, reflexed, almost as long as sepals.
Lip oblong, marked with dark brown; basal half of lip concave, with 2
broad rounded recurved side-lobes; margins undulate above.
C. Nepal to Bhutan. 900–1800 m. Oct.–Dec. Pl. 118.

526 **Coelogyne punctulata** Lindley (*C. nitida* (Roxb.) Hook.f.)
Pseudobulbs 2.5–7.5 cm long, shining. Leaves lanceolate, stalked. Flowers white, in erect or drooping clusters. Sepals *c.* 4 cm, oblong-lanceolate,
3–5–nerved. Petals blunt. Lip with rounded side-lobes marked with
yellow and brown; mid-lobe white, unmarked, ovate blunt.
E. Nepal to S.W. China. 900–2100 m. May. Pl. 115.

527 **Cymbidium aloifolium** (L.) Swartz
Epiphyte forming large pendulous crowded clusters, with pseudobulbs.
Leaves linear-oblong, to 60 cm long, notched at apex. Flowers 3.5–5 cm
across, in pendulous racemes somewhat shorter than leaves. Sepals and
petals yellowish, with faint purple stripe. Sepals oblong-lanceolate, the
lateral pair sickle-shaped. Petals ovate-oblanceolate blunt. Lip purple,

3-lobed, with apical lobe much decurved.

C. Nepal to Sikkim. 300-600 m. (India, Burma) Apr.-Jun. **Pl. 115.**

528 **Cymbidium hookerianum** See F.H. p. 403, p. 514. **Pl. 118.**

529 **Dendrobium chrysanthum** Wallich ex Lindley
Stem pendulous, 30-60 cm. Leaves lanceolate long-pointed. Flowers rather fleshy, golden yellow except for lip marked with 2 round brownish-purple spots, in clusters of 2-4 on short stalks from leafy or leafless stem. Sepals fleshy, broadly ovate, concave, keeled; lateral pair slightly sickle-shaped. Petals ovate-rounded, concave, entire or minutely toothed, slightly longer than sepals. Lip oblong-rounded or kidney-shaped, with fringed margins. The plant illustrated has unusually small flowers.

C. Nepal to Sikkim. 1300-1800 m. May-Oct. **Pl. 117.**

530 **Dendrobium denneanum** Kerr (*D. clavatum* Wallich ex Lindley)
Stem 44-75 cm. Leaves leathery, narrowly oblong, with slightly notched apex. Flowers dark yellow, 5-7.5 cm across, in lax few-flowered horizontal or decurved cluster, from leafless stem of previous year. Floral bracts oblong acute, papery. Sepals oblong, acute or blunt. Petals broadly ovate acute. Lip inrolled, with large purple blotch near centre, and fringed wavy margins.

Uttar Pradesh to Sikkim. 1200-1800 m. (Burma) May-Jun. **Pl. 116.**

531 **Dendrobium eriiflorum** Griffith
Stems erect, to 15 cm, with small pseudobulbs at base. Leaves linear-lanceolate, with oblique apex. Flowers greenish-white with purplish lip, *c.* 1.5 cm long and 1.25 cm across, in numerous many-flowered axillary clusters. Dorsal sepal lanceolate long-pointed; 2 lateral sepals sickle-shaped, with very broad bases. Petals linear acute. Lip concave; side-lobes with comb-like teeth; mid-lobe recurved, undulate.

C. Nepal to Bhutan. 1500-2100 m. (S.E. Asia) Jul.-Oct. **Pl. 117.**

532 **Dendrobium longicornu** Lindley
Epiphyte with pseudobulbs, and slender stems 15-30 cm. Leaves linear-lanceolate, with oblique apex. Flowers 5 cm long, pure white with orange lip, tapering at back into a long spur, solitary or in cluster of 2-3. Sepals and petals ovate-lanceolate, the sepals keeled. Lip with broad ridge running along the centre, and 3 lobes, the terminal lobe fringed.

C. Nepal to Bhutan. 1600-2500 m. (Burma) Sep.-Nov. **Pl. 115.**

533 **Epigeneium rotundatum** (Lindley) Summerh. (*Dendrobium r.* (Lindley) Hook.f.)
Pendulous epiphyte with woody rhizome, and pseudobulbs borne 7-10 cm apart. Leaves 2, elliptic-oblong, notched at apex. Flowers yellowish-brown, 3.5-4 cm across. Sepals ovate-lanceolate, spreading. Petals somewhat smaller than sepals. Lip obovate-oblong, with 2 rounded side-lobes and larger mid-lobe.

E. Nepal to Bhutan. 1800 m. (Burma) Apr.-May. **Pl. 116.**

534 **Gastrochilus calceolaris** (Smith) D. Don (*Saccolabium c.* (Smith) Lindley)
Epiphyte with short pendulous stems, and without pseudobulbs. Leaves leathery, narrowly oblong or linear, apex unequally 2–lobed. Flowers 12–18 mm across, in dense rounded many-flowered clusters much shorter than leaves. Petals oblong-ovate, pale green, with roundish brown markings. Base of lip forming a wide short yellow sac, with white semicircular apical lobe.
  C. Nepal to Bhutan. 1500–2200 m. (Burma, S.E. Asia) Mar.–May. Pl. 117.

535 **Oberonia falconeri** Hook.f.
Small inconspicuous tufted epiphyte. Leaves in 2 ranks, over-lapping, shortly broadly sword-shaped, with acute apex. Flowers *c.* 1.2 mm, greenish-yellow, in dense erect or decurved cylindric spike-like cluster. Bracts ovate-lanceolate or oblong, toothed. Petals ovate. Lip longer than sepals, with minute lateral lobes and divided mid-lobe.
  Uttar Pradesh to C. Nepal. 550–800 m. Aug. Pl. 116.

536 **Oberonia rufilabris** Lindley
Similar to 535, but lip of flower reddish-brown, with thread-like lateral lobes at base, and 2 short terminal lobes with acute recurved tips. Floral bracts with long thread-like tips on lower part of flower-spike.
  C. Nepal to Sikkim. 950–1500 m. Mar.–Apr. Pl. 116.

537 **Panisea uniflora** (Lindley) Lindley (*Coelogyne u.* Lindley)
Pseudobulbs broadly ovoid, tufted. Leaves paired, narrow-oblong long-pointed, narrowed to base. Flowers pale brown, solitary, on short stalks sheathed in lanceolate long-pointed overlapping bracts. Sepals 2.5 cm, oblong-lanceolate, spreading. Petals shorter, broadly lanceolate acute. Lip spotted with yellow, oblong, with short claw at base; side-lobes erect, with acute apices pointing forwards; mid-lobe ovate blunt.
  C. Nepal to Sikkim. 150–1200 m. (Burma) Apr.–May. Pl. 117.

538 **Pholidota imbricata** Hook.
Pseudobulbs cylindric, furrowed. Leaf solitary, elliptic-lanceolate acute, tapering to stout leaf-stalk. Flowers brownish- or pinkish-white, in drooping clusters 7.5–20 cm long borne on slender stems. Floral bracts large, convolute. Sepals *c.* 6 mm; dorsal sepal orbicular; lateral sepals fused at base, boat-shaped, with winged keel. Petals oblong acute. Lip 3–lobed, with mid-lobe divided at apex.
  Uttar Pradesh to S.W. China. 600–2900 m. (India, Burma, S.E. Asia, Australia) May–Aug. Pl. 116.

539 **Smitinandia micrantha** (Lindley) Lindley (*Saccolabium m.* Lindley, *Cleisostoma m.* King & Pantling)
Epiphyte with stout stem covered with leaf-sheaths. Leaves narrow-oblong, apex cut-off and slightly notched, base shortly sheathed. Flowers

purplish-pink, *c*. 3 mm across, in many-flowered spike-like clusters 5–10 cm long. Sepals broadly ovate, spreading. Petals smaller, oblong, spreading. Lip fleshy, as long as sepals, with wide blunt spur at base; apical lobe oblong, 2 lateral lobes small.

Uttar Pradesh to Bhutan. To 1100 m. (India) Apr.–May. Pl. 117.

ZINGIBERACEAE   Ginger Family

540 **Amomum subulatum** Roxb.   GREATER CARDAMOM
Perennial with leafy stem to 1½ m. Leaves oblong-lanceolate, 30–60 cm, hairless. Flowers pale yellow, 2.5 cm long, in dense globose clusters direct from stout rootstock. Bracts red-brown; outer bracts with long rigid horny tips. Lip of corolla obovate-wedge-shaped, notched, rather longer than the corolla-segments. Capsule red-brown, globose, covered with bristles. Used as a spice. Cultivated in damp shady places, often beneath alders.

C. Nepal to Sikkim. 1000–2000 m. May–Jun. Pl. 120.
Despite some printed references to the occurrence of *Elettaria cardamom* LESSER or TRUE CARDAMOM in Nepal, there seem to be no authenticated records. It is native of and cultivated in S. India.

541 **Cautleya gracilis** See F.H. p. 409. Pl. 120.

542 **Cautleya spicata** See F.H. p. 409, p. 514. Pl. 119.

543 **Hedychium densiflorum** Wallich
Leaves oblong long-pointed, to 30 cm long, hairless beneath. Flowers orange-cream, in long erect dense spike. Bracts small, oblong, rolled tightly around calyx 2.5 cm long. Corolla-tube 2.5–3.75 cm long. Staminodes lanceolate. Lip wedge-shaped, deeply 2–lobed. Stamen as long as lip, with orange-yellow filament and linear anther.

C. Nepal to Arunachal Pradesh. 1800–2800 m. Jul.–Aug. Pl. 119.

544 **Hedychium gardnerianum** Sheppard ex Ker-Gawler
Stem leafy, to 150 cm. Leaves oblong long-pointed, white-powdered beneath. Flowers bright lemon-yellow, in erect dense spike 20–45 cm long. Bracts large, oblong, hairless, white-powdered when young, as long or longer than calyces. Corolla-tube a little longer than bract, with reflexed segments. Staminodes oblanceolate, 2.5 cm long. Lip 2.5 cm long or more, obovate-wedge-shaped, 2–3–toothed, narrowed to short claw. Stamen twice length of lip, with bright red filament and linear anther.

C. Nepal to Sikkim. 1900 m. (Widely cultivated in many countries) Aug.–Sep. Pl. 119.

545 **Hedychium thyrsiforme** Buch.-Ham. ex Smith
Stem erect, leafy, to 120 cm. Leaves oblong-lanceolate, long-tailed at apex, finely hairy beneath. Flowers white, in dense oblong spike, each flower subtended by green cylindric bract. Calyx tubular, 3–toothed.

Corolla-tube hardly exceeding bract; corolla-segments linear, 2.5 cm long. Staminodes linear, as long as segments. Lip clawed, with 2 linear-oblong lobes. Stamen twice as long as lip, with white filament and linear anther. Uttar Pradesh to Sikkim. To 1200 m. Sep. Pl. 119.

546 Kaempferia rotunda L.

Perennial, leafless when in flower. Leaves erect, short-stalked, orbicular-oblong, variegated dark and light green above, purple beneath. Flowers in spikes direct from tuberous rootstock. Bracts oblong acute. Calyx slit down one side, minutely toothed. Corolla-tube 5–7.5 cm long; corolla-segments nearly as long, linear, spreading. Staminodes white, oblong acute, to 5 cm long. Lip reddish or lilac, cut into 2 rounded reflexed lobes. Anther-crest cut into 2 lanceolate lobes. This species has medicinal uses. In Nepal it has been recorded only from cultivated areas, so perhaps it is not truly native there.

C. & E. Nepal. 1500–2100 m. (India, S.E. Asia) Apr.–May. Pl. 122.

547 Roscoea purpurea See F.H. p. 408.

Further work has recently been done on *Roscoea*. See E.J. Cowley (1982), *Kew Bull.* 36, p. 747. Material previously lumped as one species has now been divided into two. The plant named on Pl. 119 of F.H. as *Roscoea purpurea* should now be named *Roscoea auriculata*. The two species are distinguishable as follows:

Roscoea auriculata Schumann

All leaves clearly eared at base. Flowers usually bright purple or white with white staminodes. Staminodes with short claw. Lip deflexed. Anther with stout connective elongation. Epigynous glands (i.e. on pistil) 4–5.5 mm long.

E. Nepal to Sikkim. 2100–4900 m. May.–Sep.

Roscoea purpurea Smith

Lower leaves sometimes slightly eared at base. Flowers usually pale lilac to mauve, with longer narrower floral segments. Lip not deflexed. Anther with long connective elongation. Epigynous glands 7.5–9.5 mm long.

Himachal Pradesh to Bhutan. 1500–3100 m. Jun.–Sep. Pl. 120.

548 Zingiber chrysanthum Roscoe

Perennial with leafy stem to 2 m. Leaves oblong-lanceolate, more or less downy beneath. Flowers in dense globose or oblong head direct from tuberous rootstock. Bracts green, the outer bracts ovate, the inner lanceolate and with sharp hairy tip. Corolla-tube 3.75–5 cm long; segments bright red, lanceolate. Lip bright yellow, 2.5 cm, deeply 3–lobed. Capsule bright red, oblong, splitting open when ripe. Seed as large as a pea, globose, with white translucent aril.

Uttar Pradesh to Sikkim. 450–1600 m. Aug. Fruiting Oct. Pl. 120.

IRIDACEAE Iris Family

549 **Belamcanda chinensis** See F.H. p. 410, p. 514. **Pl. 121.**

550 **Iris clarkei** See F.H. p. 411. **Pl. 121.**

551 **Iris germanica** See F.H. p. 413.
This iris is commonly grown in Kashmir on Muslim graves. The flowers vary from deep to pale mauve, but there is also a widespread form with white flowers. This white form resembles 552, but is distinguished by its shorter (to 5 cm), broader floral bracts, which are brown-papery for half their length or more. **Pl. 121.**

552 **Iris kashmiriana** Baker
Flowers white, sometimes tinged with pale blue; falls with a dense beard of white yellow-tipped hairs. Floral bracts green, with only a narrow papery margin, longer (7–11 cm) and narrower than the bracts of 551. Not nearly so common in the Kashmir valley as 551. It seems to be recorded only from the vicinity of cultivated areas.
Kashmir. 1500–1800 m. (Recorded from W. Asia, but probably not native there) Apr. **Pl. 121.**

553 **Iris potaninii** Maxim.
Small rhizomatous perennial, with base of stem surrounded with the dense fibrous thread-like remains of old leaves. Basal leaves linear acute, about as long as the flower-stem; stem-leaves 2–3, with awl-shaped tip and papery basal sheath. Flowers solitary, deep blue or purple. Falls bearded, *c.* 3 cm long, with obovate round-tipped limb *c.* 1 cm broad, narrowed into a claw. Standards 2.5 cm long by 0.5 cm broad, sharply notched at apex. Not recorded from our area, but it may perhaps occur there, for it is quite abundant in Tibet on the Thong La about 50 km north of the Nepal border.
Tibet to Siberia. 4000–5200 m. May–Jun. **Pl. 122.**

AMARYLLIDACEAE Daffodil Family

554 **Allium semenovii** See F.H. p. 414, p. 515. **Pl. 123.**

DIOSCOREACEAE Yam Family

555 **Dioscorea bulbifera** L.
Hairless climber with large tuberous edible roots, and with stems bearing numerous brown angular warted bulbils. Leaves broadly ovate-cordate, long-tailed, stalked. Flowers very fragrant, white or yellowish-green fading to dull red. Male flowers in slender unbranched clustered spikes

2.5–10 cm long. Female flowers in pendulous clusters 10–25 cm long. Capsule 3–winged. Seed winged on one side only. Cultivated, and also indigenous. The yams of the wild form can be eaten only after considerable preparation.

Pakistan to Bhutan. To 2100 m. (Tropics of the Old World) Jun.–Aug. **Pl. 123.**

556 **Dioscorea deltoidea** Wallich ex Griseb.

Large climbing herb with twining stem. Leaves ovate long-pointed, often as broad as long, base cordate and with rounded lobes, stalked. Male flowers with tubular perianth and short spreading lobes, in long slender often branching spikes which form a lax cluster. Female flowers smaller, with petals free, in shorter solitary spikes. Fruit a broad 3–winged capsule. Seed winged all round. The thick fleshy tuberous rootstock is not edible, but is used for washing clothes.

Afghanistan to S.W. China. 450–3100 m. May–Aug. **Pl. 123.**

LILIACEAE   Lily Family

557 **Campylandra aurantiaca**   See F.H. p. 420, p. 515. **Pl. 122.**

558 **Chlorophytum nepalense**   See F.H. p. 420, p. 516. **Pl. 123.**

559 **Fritillaria cirrhosa**   See F.H. p. 422. **Pl. 124.**

560 **Fritillaria imperialis**   CROWN IMPERIAL   See F.H. p. 422.

The centre of distribution of this species is in Iran and Afghanistan. It has been in cultivation for many centuries, and is grown in gardens and on Muslim graves in the Kashmir valley. It also occurs there in a naturalized state on steep rocky slopes surrounding the valley. In view of the absence of records of its occurrence in any locality other than ones quite close to cultivated areas, it must remain doubtful whether this species is truly indigenous. **Pl. 123.**

561 **Lilium nanum**   See F.H. p. 423, p. 517. **Pl. 124.**

562 **Lloydia flavonutans**   See F.H. p. 425. **Pl. 122.**

563 **Lloydia longiscapa**   See F.H. p. 425, p. 517. **Pl. 122.**

564 **Polygonatum oppositifolium** (Wallich) Royle

Forest perennial with creeping rootstock and erect furrowed stem, often epiphytic. Leaves opposite, ovate-lanceolate long-pointed, roundish at base, shining above, short-stalked. Flowers greenish-white, 12–15 mm long, in 3–8–flowered axillary clusters. Corolla-tube cylindric, with triangular lobes. Berry scarlet, ovoid-elliptic.

C. Nepal to Sikkim. 1800–2100 m. Apr.–May. **Pl. 122.**

565 **Polygonatum verticillatum**   WHORLED SOLOMON'S SEAL   See F.H. p. 426. **Pl. 126.**

566 **Smilax aspera** See F.H. p. 427. **Pl. 126.**

567 **Smilax ferox** See F.H. p. 428, p. 516. **Pl. 125.**

568 **Smilax orthoptera** A.DC.
Large climber with prickly branches. Leaves ovate-oblong or elliptic acute, blue-green, 5–nerved, with winged leaf-stalk. Flowers greenish-white, 5 mm long, in long-stalked umbels 2–3 cm across; umbels solitary or 2–3 together. Sepals linear-oblong. Berry 8 mm across.
C. Nepal to Sikkim. 1500–1800 m. Feb.–Mar. **Pl. 125.**

569 **Smilax rigida** Wallich ex Kunth
Rigid prickly much-branched forest undershrub, with acutely angled densely leafy spreading branches. Leaves orbicular-ovate, base more or less cordate, faintly 3–veined, leathery, short-stalked or stalkless. Flowers pinkish- or brownish-green, 2–3 mm across, in 1–4–flowered umbels. Male perianth-segments 6, very small. Female perianth-segments rather larger, with 3 staminodes. Berry black, globose.
C. Nepal to Bhutan. 2100–2900 m. May. **Pl. 125.**

570 **Tulipa stellata** See F.H. p. 429.
The plant illustrated here, with white petals with broad red bands on the outside, occurs as a common cornfield weed in the W. Himalaya. The smaller plant with red and yellow flowers shown on Pl. 124 of F.H. is var. *chrysantha*. **Pl. 124.**

PONTEDERIACEAE

571 **Monochoria vaginalis** (Burm.f.) C. Presl
Water-plant of rice-fields and irrigation-channels. Leaves fleshy, varying in shape from linear to ovate-cordate long-pointed, sheathed at base, long-stalked. Flowers blue, in a short dense spike from sheath of uppermost leaf. Perianth campanulate, 6–10 mm long, with 6 lobes. Stamens 6; one large, with blue toothed filament and blue anther; 5 smaller, with yellow anthers.
Kashmir to Bhutan. To 1800 m. (India, China, S.E. Asia) May–Sep. **Pl. 126.**

COMMELINACEAE Spiderwort Family

572 **Commelina paludosa** See F.H. p. 429, p. 518. **Pl. 125.**

573 **Cyanotis vaga** See F.H. p. 430, p. 518. **Pl. 125.**

JUNCACEAE   Rush Family

574   **Juncus membranaceus** Royle
Plant of damp places in Tibetan borderlands. Stems tufted, 12–50 cm, bearing one or more leaves above the middle. Basal sheaths papery, eared. Leaves thread-like, acute. Flowers white, in solitary terminal heads. Floral bracts pale yellow, ovate-lanceolate. Sepals *c.* 8 mm long, oblong-lanceolate blunt, papery. Anthers exserted. Capsule long-beaked, seed long-tailed.
Pakistan to Sikkim. 3000–4300 m. Jul.–Aug. Pl. 124.

PALMAE   Palm Family

575   **Phoenix humilis** Royle ex Baccari & Hook.f.
Palm with stem 1.5–6 m densely covered with stumps of old leaf-stalks. Leaves pinnate; leaflets thin, smooth, persistently folded, faintly nerved, somewhat scattered. Male flowers creamy white, fragrant, 4–5 mm long, borne on spadix *c.* 30 cm. Female flowers yellowish-green, globose. Fruit red when ripe, oblong, borne almost horizontally in spikes with stems 50–62 cm. Growth is slow, and for some years the stem is scarcely raised above the ground, in which state it much resembles *Phoenis acaulis*. This latter species can be distinguished by its thick, strongly nerved leaves which are folded at first but which open out when mature; by its shorter fruiting spadix 15–20 cm, which is often concealed in the sheaths of leaf-stalks; and by its fruit blackish when ripe, seated on the spikelets at a narrow vertical angle.
Uttar Pradesh to S.W. China. To 900 m. (India, Burma, S.E. Asia) Oct.–Nov. Fruiting Jun.–Jul. Pl. 127.

576   **Phoenix sylvestris** WILD DATE PALM   See F.H. p. 431. Pl. 127.

577   **Trachycarpus martianus** (Wallich) Wendl.
Slender palm with naked trunk to 15 m, clothed beneath the crown with persistent leaf-sheaths. Leaves fan-shaped, suborbicular, rigid, cut very regularly to about half-way down into many 2–lobed segments. Inflorescence stout, branched. Fruit bluish or yellowish, *c.* 12 mm, oblong, rounded at both ends. *Trachycarpus takil* of Uttar Pradesh and W. Nepal is somewhat similar, but is distinguished by its trunk covered with fibrous hairs, its leaf irregularly cut, and its kidney-shaped fruit. See F.H. p. 432.
C. Nepal eastwards. Planted in Kathmandu and at other places, but probably also indigenous. 1500–2000 m. Pl. 127.

70

ARACEAE   Arum Family

578   Amorphophallus bulbifer (Schott) Blume
Perennial, leafless when in flower, with globose tuberous root. Leaves with 3 pinnately cut segments, ultimately bearing bulbils in forks and on nerves above; leaf-stalk mottled grey and white. Spathe erect, 12–20 cm, mottled pink and white outside, rose-pink inside, with broad inflated tube and longer boat-shaped limb. Spadix not longer than spathe, with a dense-flowered inflorescence; appendage as long as inflorescence, flesh-coloured, rounded at top.

C. Nepal to Sikkim. 300–900 m. (India, Burma) May–Jun. **Pl. 128.**

579   Colocasia fallax Schott
Tuberous perennial, flowering when in leaf. Leaves peltate, on long leaf-stalk to 35 cm; blade orbicular-ovate, strongly nerved, base cordate or notched. Tube of spathe green, much shorter than the cream-coloured lanceolate reflexed limb. Spadix cream-coloured, 6–8 cm long, with narrow neck between male and female flowers, and with slender cylindric appendage. Anthers with rounded teeth which radiate like the points of a star.

C. Nepal to Sikkim. 400–2000 m. May–Aug. **Pl. 128.**

580   Gonatanthus pumilus   See F.H. p. 438, p. 518. **Pl. 128.**

581   Rhaphidophora decursiva (Roxb.) Schott
Robust evergreen climber, ascending trees to 10 m or more by means of roots produced from stem. Leaves large, glossy, bright green on both surfaces, in 2 ranks, pinnately lobed; lobes 8–15 pairs, 1–ribbed, with obliquely cut acute tips; leaf-stalk stout, abruptly bent at tip. Spathe pale yellow, 12.5–17.5 cm, leathery, soon falling. Spadix shorter, white, cylindric, with crowded flowers.

Uttar Pradesh to S.W. China. To 1500 m. (Burma, S.E. Asia) Nov. **Pl. 128.**

GRAMINEAE   Grass Family

582   Cymbopogon flexuosus (Nees ex Steudel) W. Watson   LEMON-GRASS
Perennial to 3 m tall, with short thick rhizome. Leaf-blade linear long-pointed, with long thread-like tip, rough on the margins and outer nerves. Inflorescence with many long drooping flexuous branches which subdivide and end in paired clusters of small stalkless spikelets. Cultivated in parts of India for a valuable aromatic oil known as 'oil of lemon-grass'.

Uttar Pradesh to Bhutan. To 1500 m. (India, Burma, S.E. Asia) Sep.–Nov. **Pl. 126.**

583   Thysanolaena maxima (Roxb.) Kuntze
Elegant perennial to 2 m, with solid reed-like culms, forming large

clumps. Leaf-blade very broad. Spikelets all similar, minute, stalked, in large branched much-divided clusters. Indigenous to our area, also grown at edges of fields. The flowering stems are cut and bound together for use as brooms.

Pakistan to Sikkim. To 1800 m. (India, S.E. Asia) Mar.–Jun. **Pl. 126.**

POLYPODIACEAE

584 **Cyathea spinulosa** Wallich ex Hook.

Very spiny tree-fern to 4 m, in areas of high rainfall. This is the common tree-fern of Nepal. From nearby Sikkim nine different species of tree-fern have been recorded, in the genera *Alsophila, Cyathea* and *Hemitelia*. See P.N. Mehra and S.S. Bir (1964), *Research Bull. Panjab Univ.*, Vol. 15, pts 1 & 2, p. 69–182. Specimens of tree-ferns are not easy to handle and in consequence they tend to be under-collected. It seems likely that at least some of the Sikkim species also occur in Nepal.

C. Nepal eastwards. To 2250 m. **Pl. 128.**

8 *Thalictrum foliolosum*

7 *Anemone rupestris*

1

Aconitum gammiei

3 *Aconitum naviculare*

3 *Delphinium glaciale*

14 *Delphinium nepalense*

4 *Adonis aestivalis*   5 *Anemone biflora*   8 *Anemone vitifolia*

2

15 *Delphinium stapeliosum*   6 *Anemone rivularis*

Clematis alternata

11 Clematis buchananiana

9 Clematis acuminata

3

Aconitum bhedingense

12 Clematis tibetana

23 *Berberis asiatica*

24 *Berberis erythroclada*

28 *Berberis mucrifolia*

4

27 *Berberis lycium*

25 *Berberis insignis*

6 *Ranunculus diffusus* 26 *Berberis koehniana* 39 *Barbarea vulgaris*

5

2 *Cocculus laurifolius* 51 *Casearia glomerata*

19 *Dillenia pentagyna*

20 *Magnolia campbellii*

21 *Michelia kisopa*

6

92 *Lannea coromandelica*

17 *Ranunculus tricuspis*

0 *Corydalis alburyi*

1 *Corydalis gerdae*

2 *Corydalis latiflora*

35 *Meconopsis napaulensis*

34 *Meconopsis bella*

33 *Corydalis megacalyx*

47 *Pegaeophyton scapiflorum*

36 *Arabidopsis himalaica*

8

90 *Cotinius coggygria*

29 *Argemone mexicana*

Arcyosperma primulifolium

Erysimum chamaephyton

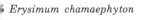

44 Ermania himalayensis

9

48 Staintoniella nepalensis

Crambe kotschyana

42 *Cardamine violacea* 49 *Thlaspi cochleariforme* 37 *Arabis pterosperma*

10

46 *Lignariella hobsonii* 41 *Cardamine macrophylla*

0 *Silene nigrescens*

6 *Arenaria polytrichoides*

57 *Silene fissicalyx*

11

59 *Silene longicarpophora*

Silene gonosperma

54 *Arenaria festucoides*

55 *Arenaria globiflora*

12

62 *Stellaria congestiflora*

53 *Arenaria edgeworthiana*

Silene tenuis   70 Abelmoschus manihot   65 Hypericum hookerianum

2 Polygala arillata   52 Polygala arillata

50 *Capparis zeylanica*

72 *Bombax ceiba*

14

66 *Hypericum podocarpoides*

71 *Hibiscus trionum*

67 *Hypericum uralum*

64 *Hypericum cordifolium*

Bombax ceiba

63 *Myricaria elegans*

15

Eurya acuminata

69 *Eurya cerasifolia*

73 *Abroma angustifolia*

78 *Geranium pratense*

16

76 *Hiptage benghalensis*

75 *Grewia optiva*

Geranium procurrens

Reevesia sp.

77 *Geranium lambertii*

17

81 *Oxalis latifolia*

Oxalis acetosella

84 *Ilex excelsa*

87 *Euonymus pendulus*

18

85 *Euonymus fimbriatus*

86 *Euonymus grandiflorus*

3 *Toona serrata*

89 *Cissus javana*

19

8 *Zizyphus mauritiana*

82 *Micromelum integerrimum*

91 *Dobinea vulgaris*

94 *Coriaria napalensis*

95 *Moringa oleifera*

20

93 *Rhus wallichii*

94 *Coriaria napalensis*

Bauhinia variegata

9 Lathyrus laevigatus

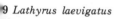

Cassia floribunda

96 Bauhinia variegata

21

98 Caesalpinia cucculata

102 *Astragalus frigidus*

100 *Astragalus chlorostachys*

22

101 *Astragalus donianus*

104 *Astragalus subuliformis*

Campylotropis speciosa

Butea monosperma

103 *Astragalus melanostachys*

23

107 *Caragana jubata*

105 *Butea monosperma*

108 *Caragana sukiensis*   99 *Apios carnea*   109 *Cochlianthus gracilis*

24

110 *Crotalaria cytisoides*   111 *Crotalaria tetragona*

2 *Derris microptera*

116 *Hedysarum manaslense*

7 *Indigofera exilis*

118 *Indigofera pulchella*

114 *Desmodium multiflorum*

113 *Desmodium confertum*

26

121 *Mucuna macrocarpa*

115 *Hedysarum kumaonense*

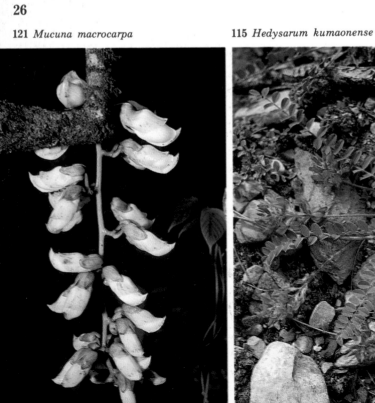

26 *Oxytropis mollis*

24 *Oxytropis humifusa*

120 *Lotus corniculatus*

27

127 *Thermopsis inflata*

128 *Thermopsis lanceolata*

122 *Oxytropis arenae-ripariae*

125 *Oxytropis microphylla*

123A *Oxytropis cachemiriana*

28

123B *Oxytropis cachemiriana*

129 *Trigonella emodi*

138 *Potentilla plurijuga*

132 *Cotoneaster integrifolius*

137 *Potentilla atrosanguinea*

29

135 *Maddenia himalaica*

136 *Photinia integrifolia*

133 *Cotoneaster roseus*

134 *Eriobotrya elliptica*

140 *Prunus napaulensis*

30

131 *Cotoneaster frigidus*

130 *Cotoneaster acuminatus*

147 *Rubus niveus*

144 *Rubus foliolosus*

141 *Prunus prostrata*

31

148 *Rubus splendidissimus*

145 *Rubus hypargyrus*

141 *Prunus prostrata* 142 *Pyracantha crenulata*

32

146 *Rubus nepalensis* 143 *Rubus calycinus*

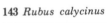

139 *Prunus bokhariensis* 139 *Prunus bokhariensis*

3 Sorbus microphylla

0 Sorbus foliolosa

151 *Sorbus insignis*

33

155 *Sorbus wallichii*

4 Sorbus ursina

149 *Sorbus cuspidata*

152 *Sorbus lanata*

34

169 *Hydrangea aspera*

156 *Astilbe rivularis*

170 *Ribes alpestre*

172 *Ribes himalense*

4 *Ribes orientale*

3 *Ribes laciniatum*

171 *Ribes glaciale*

167 *Saxifraga pilifera*

165 *Saxifraga melanocentra*

163 *Saxifraga hirculoides*

36

166 *Saxifraga nutans*

161 *Saxifraga caveana*

164 *Saxifraga hypostoma*

162 *Saxifraga hemisphaerica*

161 *Saxifraga caveana*

168 *Saxifraga punctulata*

157 *Bergenia purpurascens*

160 *Saxifraga brachypoda*

159 *Saxifraga aristulata*

178 *Rosularia alpestris*

38

180 *Sedum multicaule*

158 *Chrysosplenium forrestii*

7 *Rhodiola fastigiata*

176 *Rhodiola cretinii*

5 *Rhodiola amabilis*

179 *Sedum filipes*

81 *Sedum oreades*

181 *Sedum oreades*

182 *Combretum roxburghii*

185 *Duabanga grandiflora*

40

186 *Trichosanthes lepiniana*

184 *Rotala rotundifolia*

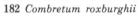

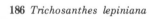

83 *Callistemon citrinus*

88 *Eryngium billardieri*

7 *Cereus peruvianus*

189 *Heracleum nepalense*

41

190 *Heracleum wallichii*

200 *Trevesia palmata*

191 *Acanthopanax cissifolius*

198 *Schefflera impressa*

42

193 *Brassaiopsis mitis*

194 *Hedera nepalensis*

92 *Aralia cachemirica*

197 *Pentapanax leschenaultii*

99 *Schefflera venulosa*

196 *Merrilliopanax alpinus*

202 *Alangium alpinum*

195 *Helwingia himalaica*

44

203 *Alangium salviifolium*

203 *Alangium salviifolium*

201 *Swida oblonga*

212 *Viburnum cylindricum*

204 *Lonicera angustifolia*

205 *Lonicera hypoleuca*

206 *Lonicera lanceolata*

45

213 *Viburnum mullaha*

208 *Lonicera tomentella*

211 *Viburnum colebrookianum*　　　209 *Sambucus adnata*

46

207 *Lonicera ligustrina*　　210 *Sambucus wightiana* Kashmir

214 *Asperula oppositifolia*

216 *Mussaenda roxburghii*

215 *Galium verum*

217 *Wendlandia coriacea*

224 *Artemisia gmelinii* 223 *Artemisia dracunculus* 222 *Artemisia absinthium*

48

218 *Ageratum conyzoides* 219 *Anaphalis busua*

216 *Aster molliusculus*

211 *Centaurea iberica*

217 *Aster sikkimensis*

221 *Arctium lappa*

49

220 *Anthemis cotula*

228 *Aster thomsonii*　　229 *Aster trinervius*　　225 *Aster indamellus*

50

233 *Cichorium intybus*　234 *Crassocephalum crepidioides* 230 *Carduus edelbergii*

238 *Cyathocline purpurea*

51

261 *Tanacetum nubigenum*

37 *Crepis sancta*

32 *Chrysanthemum pyrethroides*

236 *Crepis flexuosa*

242 *Gynura cusimbua*   243 *Gynura nepalensis*   259 *Sphaeranthus indicus*

52

235 *Cremanthodium purpureifolium*   239 *Dubyaea hispida*

241 *Gnaphalium affine*

240 *Eupatorium odoratum*

53

245 *Jurinea* sp.

244 *Jurinea ceratocarpa* var. *depressa*

246 *Leontopodium monocephalum*

263 *Waldheimia nivea*

247 *Leontopodium stracheyi*

54

249 *Nannoglottis hookeri*

264 *Youngia tenuifolia* subsp. *diversifolia*

2 *Saussurea bracteata*

253 *Saussurea costus*

8 *Soroseris pumila*

251 *Saussurea atkinsonii*

250 *Petasites tricholobus*

56

260 *Tagetes patula*

262 *Vernonia volkameriifolia*

248 *Ligularia fischeri*

257 *Senecio wallichii*

255 *Senecio diversifolius*

256 *Senecio triligulatus*

254 *Senecio cappa*

266 *Codonopsis thalictrifolia*

268 *Lobelia seguinii* var. *doniana*

58

267 *Cyananthus microphyllus*

284 *Pyrola karakoramica*

273 *Gaultheria trichophylla*

285 *Diapensia himalaica*

5 *Codonopsis purpurea*

4 *Vaccinium dunalianum*

271 *Gaultheria griffithiana*

59

270 *Gaultheria fragrantissima*

4 *Agapetes serpens*

272 *Gaultheria hookeri*

276 *Rhododendron campylocarpum*

60

283 *Vaccinium sikkimense*

282 *Vaccinium retusum*

8 *Rhododendron grande*

277 *Rhododendron fulgens*

61

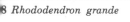4 *Rhododendron arboreum*

275 *Rhododendron campanulatum*

280 *Rhododendron pumilum*

279 *Rhododendron lanatum*

62

287 *Androsace muscoidea*

288 *Androsace muscoidea* forma *longiscapa*

286 *Androsace lanuginosa*

289 *Androsace nortonii*

1 *Primula boothii*

293 *Primula buryana* var. *purpurea*

63

2 *Primula gracilipes*

295 *Primula concinna*

Primula drummondiana

290 *Primula barnardoana*

304 *Primula minutissima*

300 *Primula glabra*

64

301 *Primula glandulifera*

305 *Primula muscoides*

313 *Primula tenuiloba*

312 *Primula soldanelloides*

Primula megalocarpa

314 *Primula tibetica*

Primula ramzanae

309 *Primula sessilis*

Primula sharmae

296 *Primula deuteronana*

315 *Primula uniflora*

297 *Primula dickieana*

66

292 *Primula buryana*

311 *Primula sibirica*

Primula capitata subsp. crispata

308 Primula reidii var. williamsii

Primula pulchra

299 Primula floribunda

318 *Maesa macrophylla*

68

319 *Myrsine semiserrata*

316 *Ardisia macrocarpa*

317 *Ardisia solanacea*

321 *Styrax hookeri*

2 *Symplocos dryophila*

325 *Symplocos ramosissima*

0 *Aesandra butyracea*

323 *Symplocos glomerata*

332 *Syringa emodi*

324 *Symplocos pyrifolia*

331 *Ligustrum indicum*

70

326 *Symplocos sumuntia*

327 *Symplocos theifolia*

Jasminum multiflorum   328 Jasminum dispermum   329 Jasminum mesneyi

71

Cynanchum auriculatum   345 Vincetoxicum hirundinaria   345 Vincetoxicum hirundinaria

341 *Ceropegia wallichii*   343 *Pentasacme wallichii*   338 *Ceropegia hookeri*

72

334 *Chonemorpha fragrans*   340 *Ceropegia pubescens*

333 *Alstonia neriifolia*

335 *Plumeria rubra*

337 *Asclepias curassavica*

336 *Wrightia arborea*

339 *Ceropegia longifolia*

344 *Raphistemma pulchellum*

349 *Gentiana elwesii*

74

351 *Gentiana stipitata*

352 *Gentiana urnula*

350 *Gentiana robusta*

355 *Swertia nervosa*

354 *Swertia multicaulis*

346 *Buddleja asiatica*

347 *Buddleja crispa*

348 *Gentiana algida*

75

353 *Swertia cuneata*

362 *Pseudomertensia racemosa*

356 *Arnebia guttata*

360 *Maharanga emodi*

76

357 *Eritrichium canum*

358 *Eritrichium nanum* subsp. *villosum*

365 *Solanum torvum*   364 *Nicandra physalodes*   363 *Porana racemosa*

361 *Microula sikkimensis*   359 *Lindelofia anchusoides*

366 *Hemiphragma heterophyllum*

366 *Hemiphragma heterophyllum*

78

381 *Veronica lanuginosa*

372 *Pedicularis elwesii*

370 *Oreosolen wattii*

376 *Pedicularis rhinanthoides*

375 *Pedicularis megalantha*

378 *Pedicularis siphonantha*

374 *Pedicularis longiflora* var. *tubiformis* Zanskar

373 *Pedicularis klotzschii*

377 *Pedicularis roylei*

383 *Veronica persica*

371 *Pedicularis cheilanthifolia*

367 *Lagotis cashmeriana*

368 *Lagotis kunawurensis*

364 *Wulfenia amherstiana*

369 *Mazus dentatus*

385 *Aeginetia indica*   386 *Orobanche aegyptiaca*   387 *Orobanche solmsii*

379 *Scrophularia*
   decomposita   382 *Veronica laxa*   380 *Veronica himalensis*

392 *Didymocarpus aromaticus*

388 *Aeschynanthus hookeri*

83

390 *Chirita pumila*

394 *Didymocarpus pedicellatus*

412 *Holmskioldia sanguinea*

407 *Caryopteris odorata*

391 *Corallodiscus lanuginosus*

389 *Chirita bifolia*

395 *Lysionotus serratus*

393 *Didymocarpus oblongus*

397 *Incarvillea younghusbandii*

396 *Incarvillea emodi*

399 *Echinacanthus attenuatus*

400 *Eranthemum pulchellum*

401 *Goldfussia nutans*   **403** *Thunbergia coccinea*   **404** *Thunbergia fragrans*

86

402 *Strobilanthes atropurpureus*   **398** *Aechmanthera gossypina*

05 *Callicarpa arborea*

409 *Clerodendrum serratum*

08 *Clerodendrum philippinum*

406 *Callicarpa macrophylla*

411 *Duranta repens*

413 *Vitex negundo*

410 *Clerodendrum viscosum*

410 *Clerodendrum viscosum*

415 *Ajuga lupulina*

421 *Glechoma tibetica*

414 *Ajuga lobata*

422 *Mentha longifolia*

419 *Dracocephalum*
   heterophyllum

416 *Anisomeles indica*

418 *Coleus barbatus*

90

419 *Dracocephalum heterophyllum* Dolpo, W. Nepal

5 *Nepeta lamiopsis*

423 *Nepeta coerulescens*

429 *Stachys melissaefolia*

6 *Nepeta longibracteata*

424 *Nepeta floccosa*

417 *Colebrookea oppositifolia*

420 *Glechoma nivalis*

428 *Pogostemon benghalensis*

92

427 *Nepeta nervosa*

430 *Thymus linearis*

Peperomia tetraphylla

Polygonum paronychioides

431 Cyathula capitata

93

432 Cyathula tomentosa

Bistorta millettii

433 *Chenopodium foliolosum*

437 *Aconogonum tortuosum*

434 *Aconogonum alpinum* Darcha, Lahul

441 *Fagopyrum esculentum*

442 *Bistorta vaccinifolia*

448 *Bistorta affinis* Miyah Nullah, Lahul

439 *Bistorta amplexicaulis*

440 *Bistorta emodi*

436 *Aconogonum molle*

435 *Aconogonum campanulatum*

46 *Rheum acuminatum*

444 *Persicaria polystachya*

48 *Rumex hastatus*

448 *Rumex hastatus*

450 *Rumex patentia* subsp. *tibeticus*

449 *Rumex nepalensis*

450 *Rumex patentia* subsp. *tibeticus*

447 *Rheum webbianum*

2 *Piper boehmeriaefolia*

453 *Actinodaphne obovata*

) *Phoebe lanceolata*

454 *Cinnamomum tamala*

463 *Dendropthoe falcata*

458 *Persea odoratissima*

100

456 *Litsea doshia*

455 *Lindera neesiana*

Trewia nudiflora

457 *Neolitsea pallens*

Hippophae rhamnoides subsp. *turkestanica*

462 *Hippophae salicifolia*

Sarcococca coriacea

474 *Sarcococca wallichii*

464 *Euphorbia helioscopia*

102

468 *Euphorbia stracheyi*

466 *Euphorbia pulcherrima*

466 *Euphorbia pulcherrima* C.Nepal

465 *Euphorbia prolifera*

Debregeasia salicifolia   **469** Mallotus philippinensis   **471** Sapium insigne

103

Euphorbia aff. sikkimensis   **469** Mallotus philippinensis

470 *Phyllanthus emblica*

475 *Boehmeria platyphylla*

104

476 *Boehmeria rugulosa*

460 *Grevillea robusta*

480 *Lecanthus peduncularis*

481 *Pilea glaberrima*

482 *Pilea umbrosa*

479 *Girardinia diversifolia*

484 *Ficus benjamina* var. *nuda*

478 *Elatostema sessile*

488 *Ficus oligodon*

489 *Ficus religiosa*

490 *Ficus semicordata*

483 *Ficus benghalensis*

486 *Ficus hederacea*

485 *Ficus glaberrima*

487 *Ficus hispida*

494 *Populus jacquemontiana* var. *glauca*   492 *Myrica esculenta*

108

493 *Betula utilis* Iswa Khola, E.Nepal

491 *Engelhardia spicata*

491 *Engelhardia spicata*

495 *Salix babylonica* and 40 *Brassica juncea* Kashmir valley

496 *Ephedra gerardiana*

496 *Ephedra gerardiana*

110

498 *Abies spectabilis*

497 *Abies densa*

505 *Tsuga dumosa*

504 *Pinus wallichiana*

512 *Podocarpus neriifolius*

506 *Cupressus torulosa*

509 *Juniperus macropoda*

111

509 *Juniperus macropoda*

501 *Larix himalaica*

507 *Juniperus communis*   502 *Picea smithiana*   500 *Larix griffithiana*

112

499 *Cedrus deodara*   513 *Taxus baccata* subsp. *wallichiana*

0 *Juniperus recurva*

503 *Pinus roxburghii*

1 *Juniperus squamata*

508 *Juniperus indica*

515 *Eulophia herbacea* 518 *Habenaria stenopetala* 521 *Platanthera arcuata*

114

514 *Coeloglossum viride* 520 *Neottianthe secundiflora* 519 *Neottianthe calcicola*

516 *Galearis spathulata*

532 *Dendrobium longicornu*

526 *Coelogyne punctulata*

524 *Coelogyne cristata*

522 *Spathoglottis ixioides*

527 *Cymbidium aloifolium*

533 *Epigeneium rotundatum*

538 *Pholidota imbricata*

536 *Oberonia rufilabris*

116

530 *Dendrobium denneanum*

535 *Oberonia falconeri*

534 *Gastrochilus calceolaris*

539 *Smitinandia micrantha*

537 *Panisea uniflora*

117

529 *Dendrobium chrysanthum*

531 *Dendrobium eriiflorum*

517 *Goodyera fusca*

525 *Coelogyne fuscescens*

118

528 *Cymbidium hookerianum*

523 *Aerides odorata*

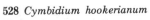

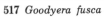

543 *Hedychium densiflorum*

542 *Cautleya spicata*

545 *Hedychium thyrsiforme*

544 *Hedychium gardnerianum*

541 *Cautleya gracilis*

547 *Roscoea purpurea*

120

540 *Amomum subulatum*

548 *Zingiber chrysanthum*

549 *Belamcanda chinensis*

550 *Iris clarkei*

552 *Iris kashmiriana*

551 *Iris germanica*

553 *Iris potaninii*

546 *Kaempferia rotunda*

122

557 *Campylandra aurantiaca*

564 *Polygonatum oppositifolium*

562 *Lloydia flavonutans*

563 *Lloydia longiscapa*

554 *Allium semenovii*   560 *Fritillaria imperialis*   558 *Chlorophytum nepalense*

123

555 *Dioscorea bulbifera*   556 *Dioscorea deltoidea*

561 *Lilium nanum*

574 *Juncus membranaceus*

124

570 *Tulipa stellata*

559 *Fritillaria cirrhosa*

569 *Smilax rigida*　　　　**567** *Smilax ferox*　　　　**568** *Smilax orthoptera*

572 *Commelina paludosa*　　　　**573** *Cyanotis vaga*

566 *Smilax aspera*   571 *Monochoria vaginalis*   565 *Polygonatum verticillatum*

126

582 *Cymbopogon flexuosus*   583 *Thysanolaena maxima*

576 *Phoenix sylvestris*

577 *Trachycarpus martianus*

575 *Phoenix humilis*

575 *Phoenix humilis*

578 *Amorphophallus*
    bulbifer

579 *Colocasia fallax*

580 *Gonatanthus pumilus*

581 *Raphidophora decursiva*

584 *Cyathea spinulosa*

Index

Accepted names are given in roman type, synonyms in italic. Numbers given are running numbers, colour plates are indicated by pl., page numbers by p.